ISNM 101:
International Series of Numerical Mathematics
Internationale Schriftenreihe zur Numerischen Mathematik
Série Internationale d'Analyse Numérique
Vol. 101

Edited by
K.-H. Hoffmann, Augsburg; H. D. Mittelmann, Tempe;
J. Todd, Pasadena

Birkhäuser Verlag
Basel · Boston · Berlin

Unilateral Problems in Structural Analysis IV

Proceedings of the fourth meeting on Unilateral Problems in Structural Analysis, Capri, June 14–16, 1989

Edited by

G. Del Piero
F. Maceri

1991

Birkhäuser Verlag
Basel · Boston · Berlin

Editors

Prof. G. Del Piero
Instituto di Meccanica Teorica
ed Applicata
Università di Udine
V.le Ungheria 43
I–33100 Udine

Prof. Ing. F. Maceri
Dipartimento di Ingegneria Civile
Università di Roma »Tor Vergata«
Via della Ricerca Scientifica
I–00173 Roma

Library of Congress Cataloging-in-Publication Data

Meeting on Unilateral Problems in Structural Analysis (4th : 1989 : Capri, Italy)
Unilateral problems in structural analysis IV : proceedings of the Fourth Meeting on Unilateral Problems in Structural Analysis, Capri, June 14–16, 1989 / edited by G. Del Piero, F. Maceri.
(International series of numerical mathematics ; v. 101)
ISBN 3-7643-2487-2 – ISBN 0-8176-2487-2
1. Structural analysis (Engineering)–Congresses. 2. Engineering mathematics–Congresses. I. Del Piero, G. (Gianpietro) II. Maceri, F. (Franco) III. Title. IV. Series.
TA646.M43 1989
624.1'71 – dc20

Deutsche Bibliothek Cataloging-in-Publication Data

Unilateral problems in structural analysis IV: proceedings of the 4. Meeting on Unilateral Problems in Structural Analysis, Capri, June 14–16, 1989 / ed. by G. del Piero; F. Maceri. – Basel ; Boston ; Berlin : Birkhäuser, 1991
(International series of numerical mathematics ; Vol. 101)
ISBN 3-7643-2487-2 (Basel ...)
ISBN 0-8176-2487-2 (Boston)
NE: Del Piero, Gianpietro [Hrsg.]; Meeting on Unilateral Problems in Structural Analysis <04, 1989, Capri>; GT

Printed from the authors' camera-ready manuscripts on acid-free paper in Germany
ISBN 3-7643-2487-2
ISBN 0-8176-2487-2

PREFACE

The present Volume contains the contributions to the fourth meeting on Unilateral Problems in Structural Analysis, held at Capri on June 14 to 16, 1989. The preceding meetings took place at Villa Emma, near Udine, on May 1982, at Ravello on September 1983 and again at Villa Emma on June 1985. Publication of the proceedings started with the second meeting; the two resulting volumes were published by Springer Verlag, Vienna, under the series Cism Courses and Lectures.

Unilateral Problems appear as a singular example of confluence of interests: they are the object of the attention of pure and applied mathematicians, of specialists in Continuum Mechanics and engineers. The idea which gave origin to this series of meetings was that of putting together people coming from such different fields. The result was an extremely fruitful exchange of experiences; it contributed, we believe, to the improvement of the knowledge in the area.

The contents of the present Volume reflects the composite character of the meeting. There are contributions in the mathematical theory (Haslinger, Panagiotopoulos, Romano), and studies in classical problems of Mechanics such as unilateral contact with friction (Kalker, Klarbring, Licht, Telega), Plasticity (Corradi, Del Piero, Owen) and composite materials and structures (Bruno, Leonardi). Some contributions deal with not yet completely explored questions of unilateral dynamics (Guo, Jean); finally, a contribution (Bennati) concerns the comparatively new subject of masonry structures, in which the unilateral constraint enters at the constitutive level.

All these fields are still in full development: many problems which were considered as open not long ago are now finding their solution, and many new ones are arising. We hope that the contributions presented here succeed in supplying a good account of the present stage of this development.

Gianpietro Del Piero
Faculty of Engineering
University of Udine
Italy

Franco Maceri
Faculty of Engineering
University of Rome "Tor Vergata"
Italy

Contents

International Series of Numerical Mathematics, Vol. 101, © 1991 Birkhäuser Verlag Basel

ELEMENTARY SOLUTIONS FOR EQUILIBRIUM PROBLEMS OF MASONRY-LIKE MATERIALS

S. Bennati
Istituto di
Scienza delle Costruzioni
Universita' di Pisa

M. Lucchesi
CNUCE
Istituto del C.N.R.
Pisa

1. Introduction

Certain features of the mechanical behaviour of materials such as masonry, which can withstand only low tension, are still far from being fully understood. Various authors have proposed that a non-linear elastic constitutive law should be used for such materials, with a negative semi-definite stress tensor and with the strain being considered as the sum of an elastic part and an inelastic part (Romano & Romano [1], Di Pasquale [2], Abruzzese et al. [3], Del Piero [4]). This constitutive law is relatively simple. However, although some progress has been made (Giaquinta & Giusti [5], Anzellotti [6], Del Piero [7]), the conditions for ensuring the existence of the solutions in a number of equilibium problems, even elementary ones, are still unknown, as are methods allowing us to determine them. Such difficulties can often be eluded by supposing the material to be linear elastic and by finding those structural shapes, if any, the interior of which is in a state of pure compression under the prescribed load conditions (Villaggio [8], Bennati & Lucchesi [9], Bennati et al. [10]). Nevertheless, it is clearly of some interest to find explicit solutions which, however elementary they may be, are obtained while still utilizing the hypothesis that the material does not support tension. On the one hand, such solutions can tell us to what extent the constitutive model is suitable for the description of the behaviour of materials such as masonry; on the other hand, they can be useful in evaluating the effectiveness of possible numerical methods [3].

Two simple problems will be examined in the present paper. The first concerns a parallelepiped-shaped block resting without friction on an elastic plate and subjected to its own weight and to a load distributed over the upper face. In the second a wedge, resting without friction on a cylindrical-shaped elastic shell, undergoes plane strain as a result of a load distributed along its edge. For both problems, we find the class of the regular (*i. e.* continuously differentiable at least three times) displacement fields which solve them. In all the cases in which a field that minimizes the inelastic strain exists, this is established. The solutions reduce to those found by Di Pasquale for the corresponding plane elements resting on a rigid base [11].

Lastly, the conditions that must verify the external loads are analyzed, so that the elastic and inelastic strain components separately satisfy the compatibility equations.

2. The constitutive law of masonry-like materials

In what follows we use $\mathbb{R}^3$, as usual, to indicate the cartesian space with coordinates (x, y, z). Moreover, if $\mathbf{A}$ and $\mathbf{B}$ are two tensors, *i.e.*, two linear mappings of $\mathbb{R}^3$ into itself, we write $\mathbf{A} \geq \mathbf{0}$ ($\mathbf{A} \leq \mathbf{0}$, respectively) to indicate that $\mathbf{A}$ is a positive (negative) semidefinite, that is that we have $\mathbf{n} \cdot \mathbf{An} \geq 0$ ($\mathbf{n} \cdot \mathbf{An} \leq 0$) for each vector $\mathbf{n} \in \mathbb{R}^3$; we shall use $\mathbf{A} \cdot \mathbf{B} = \mathrm{tr}(\mathbf{AB}^T)$, where $\mathbf{B}^T$ is the transpose of $\mathbf{B}$, to denote the inner product. Finally, we recall that if $\mathbf{A}$ and $\mathbf{B}$ are symmetric (not necessarily coaxial), one positive and the other one negative-semidefinite, then $\mathbf{A} \cdot \mathbf{B} \leq 0$.

The constitutive equation of a masonry-like material has been studied in detail by various authors (see, for example, [4] and [12]); consequently, only the essential features will be mentioned here.

The strain tensor $\mathbf{E}$ is taken to be the sum of an elastic part $\mathbf{E}^e$ and an inelastic part E^a, where E^a is positive semi-definite:

$$\mathbf{E} = \mathbf{E}^e + \mathbf{E}^a \quad , \qquad \mathbf{E}^a \geq \mathbf{0} \,. \tag{2.1}$$

Moreover, the hypothesis is made that stress $\mathbf{S}$, negative semi-definite, is orthogonal to $\mathbf{E}^a$ and that it depends linearly and isotropically on $\mathbf{E}^e$. Therefore

$$\mathbf{S} \leq \mathbf{0} \quad ; \qquad \mathbf{S} \cdot \mathbf{E}^a = \mathbf{0} \tag{2.2}$$

and

$$\mathbf{S} = 2\mu \mathbf{E}^e + (\lambda \,\mathrm{tr}\mathbf{E}^e) I \,, \tag{2.3}$$

where λ and μ denote the Lame' moduli. If it turns out that

$$\mu > 0 \quad , \qquad 2\mu + 3\lambda > 0 \,, \tag{2.4}$$

it is shown that a unique inelastic strain $\mathbf{E}^a$ and a unique stress $\mathbf{S}$ satisfying relations (2.1), (2.2) and (2.3) are associated to each strain $\mathbf{E}$ [4]. Here we suppose that condition (2.4) is satisfied and this implies that the constitutive law (2.3) is invertible. Indeed, for each symmetric tensor $\mathbf{E}$ we have

$$\mathbf{E}^e = \frac{1}{2\mu}\left(\mathbf{S} - \frac{\lambda \operatorname{tr}\mathbf{S}}{2\mu + 3\lambda}\mathbf{I}\right), \tag{2.6}$$

where $\mathbf{I}$ is the identity tensor.

It is easy to verify that, as a result of $(2.1)_2$ and (2.2), tensors $\mathbf{E}^a$ and $\mathbf{S}$ are coaxial and satisfy

$$\mathbf{S}\,\mathbf{E}^a = \mathbf{E}^a\mathbf{S} = 0 . \tag{2.7}$$

Thus, because of isotropy, $\mathbf{E}$, $\mathbf{E}^e$ and $\mathbf{S}$ are coaxial.

Lastly, it immediately follows from $(2.2)_1$ that, for each vector $\mathbf{n}$,

$$\mathbf{n}\cdot\mathbf{Sn} = \mathbf{0} \qquad \text{implies} \qquad \mathbf{Sn} = \mathbf{0} . \tag{2.8}$$

In view of the applications we are concerned with, we now prove that the constitutive law defined by relations (2.1) - (2.3) has a property that is useful in the case of plane strain. To be more precise, we wish to prove that, if we have

$$\lambda > 0 , \tag{2.9}$$

then if $\mathbf{k}$ is an eigenvector of $\mathbf{E}$ corresponding to a null eigenvalue, the eigenvalue of $\mathbf{E}^a$ corresponding to $\mathbf{k}$ is also zero.

For this purpose, let us put

$$\mathbf{Ek} = \mathbf{0} ; \tag{2.10}$$

from (2.1) we obtain

$$\mathbf{k}\cdot\mathbf{E}^e\mathbf{k} = -\,\mathbf{k}\cdot\mathbf{E}^a\mathbf{k} \le 0 . \tag{2.11}$$

Let us assume to the contrary that $\mathbf{E}^a\mathbf{k} \neq \mathbf{0}$; in view of (2.8), we have $\mathbf{k}\cdot\mathbf{E}^a\mathbf{k} \neq 0$ and taking into account (2.2) and (2.3) we get

$$2\mu(\mathbf{k}\cdot\mathbf{E}^e\mathbf{k}) + \lambda \operatorname{tr}\mathbf{E}^e = 0 . \tag{2.12}$$

On the other hand, it follows from $(2.2)_1$, $(2.4)_2$ and (2.11) that

$$\operatorname{tr}\mathbf{E}^e \le 0 \quad ; \quad \mathbf{k}\cdot\mathbf{E}^e\mathbf{k} \le 0 \tag{2.13}$$

and therefore, bearing in mind (2.9), (2.11) and (2.12), we obtain $\mathbf{k}\cdot\mathbf{E}^a\mathbf{k} = 0$, which, in view of $(2.8)_1$, implies that

$$\mathbf{E}^a\mathbf{k} = \mathbf{0} . \tag{2.14}$$

3. The case of a block resting on an elastic plate

A parallelepiped-shaped block Ω with height h, base 2b and thickness 2c, consisting of

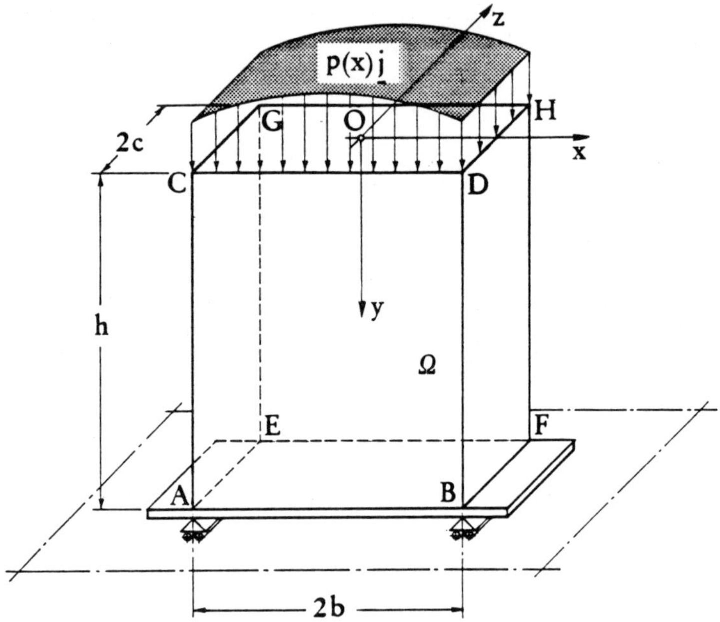

Fig. 1

masonry-like material, is resting without friction on an elastic plate which, in its turn, is resting on two supports situated below sides AE and BF of the block (Fig. 1). Besides its own weight $\gamma\mathbf{j}$, the block is subjected to load $p(x)\mathbf{j}$ on face CDHG, twice continuously differentiable, and such that

$$p(x) = p(-x) \quad , \qquad p(x) \geq 0 \; . \tag{3.1}$$

Let $x = x_1$, with $-b < x_1 < b$, be a plane that divides the block into two parts. Since $\sigma_{xy}(x,h,z) = 0$ by hypothesis, the equilibrium of either of the two parts of the block requires that

$$\int_0^h dy \int_{-c}^{+c} \sigma_{xx}(x_1,y,z)dz = 0 \; . \tag{3.2}$$

Since σ_{xx} is a non-positive function which we suppose to be continuous, given the arbitrariness of x_1, (3.2) implies that

$$\sigma_{xx}(x,y,z) = 0 \; ; \tag{3.3}$$

it, therefore, follows from (2.8) that

$$\sigma_{xy}(x,y,z) = \sigma_{xz}(x,y,z) = 0 \,. \tag{3.4}$$

In exactly the same way, it can be proved that, since $\sigma_{yz}(x,h,z) = 0$, the equilibrium of the block also implies that

$$\sigma_{zz}(x,y,z) = \sigma_{yz}(x,y,z) = 0 \,. \tag{3.5}$$

In view of (3.3)-(3.5) we may conclude that the only negative semi-definite stress field in equilibrium with the external loads is

$$\sigma_{xx} = \sigma_{zz} = \sigma_{yz} = \sigma_{xz} = \sigma_{xy} = 0 \,, \quad \sigma_{yy} = -\,[\gamma y + p(x)] \,, \tag{3.6}$$

corresponding to the elastic strains

$$\begin{aligned} \epsilon^e{}_{xx} &= \epsilon^e{}_{zz} = (\nu/E)[\gamma y + p(x)] \,, \\ \epsilon^e{}_{yy} &= -(1/E)[\gamma y + p(x)] \,, \\ \epsilon^e{}_{xy} &= \epsilon^e{}_{xz} = \epsilon^e{}_{yz} = 0 \,, \end{aligned} \tag{3.7}$$

where E and ν denote the Young modulus and the Poisson ratio:

$$E = \mu(2\mu + 3\lambda)/(\mu + \lambda) \qquad \text{and} \qquad \nu = \lambda/(\mu + \lambda) \,. \tag{3.8}$$

It then follows from the constitutive equation (2.1)-(2.3) that

$$\epsilon^a{}_{yy} = \epsilon^a{}_{xy} = \epsilon^a{}_{xz} = \epsilon^a{}_{yz} = 0 \,. \tag{3.9}$$

In agreement with equations (3.7) and (3.9), the components u(x,y) along x and v(x,y) along y of the displacement **u**, assumed to be independent of z, verify the equations

$$u_{,x}(x,y) = (\nu/E)[\gamma y + p(x)] + \varphi(x,y) \,, \tag{3.10}$$

in which the inelastic strain $\epsilon^a{}_{xx}$ is denoted by $\varphi(x,y)$, and

$$v_{,y}(x,y) = -\,(1/E)[\gamma y + p(x)] \,, \tag{3.11}$$

$$u_{,y}(x,y) + v_{,x}(x,y) = 0 \,. \tag{3.12}$$

Integrating (3.10) with respect to x and (3.11) with respect to y we obtain

$$u(x,y) = (\nu/E)[\gamma xy + P(x)] + \Phi(x,y) + C(y) , \tag{3.13}$$

$$v(x,y) = -(\gamma/2E)y^2 - (1/E)p(x)y + M(x) , \tag{3.14}$$

where $P(x)$ is a primitive of $p(x)$, $\Phi(x,y)$ is a primitive of $\varphi(x,y)$ with respect to x and $C(y)$ and $M(x)$ are arbitrary functions. On the other hand,

$$v(x,h) = -(\gamma/2E)h^2 - (1/E)p(x)h + M(x) = \eta_0(x), \tag{3.15}$$

where $\eta_0(x)$ is the solution of the differential equation

$$D_0\eta^{IV}(x) = \gamma h + p(x) , \tag{3.16}$$

in which D_0 is the flexural rigidity of the plate, satisfying the boundary conditions

$$\eta(-b) = \eta(b) =0, \qquad \eta''(-b) = \eta''(b) = 0 . \tag{3.17}$$

Thus

$$M(x) = \eta_0(x) + (1/E)hp(x) + (\gamma/2E)h^2 , \tag{3.18}$$

and

$$v(x,y) = \eta_0(x) + (\gamma/2E)(h^2 - y^2) + (1/E)p(x)(h - y) . \tag{3.19}$$

Substituting (3.13) and (3.19) in (3.12) we obtain the equation

$$(\nu\gamma/E)x + \Phi_{,y}(x,y) + C'(y) + \eta_0'(x) +$$
$$(1/E)p'(x)(h - y) = 0 , \tag{3.20}$$

from which it follows that

$$\Phi(x,y) + C(y) = - \eta_0'(x)y - (1/E)p'(x)(h -y/2)y +$$
$$- (\nu\gamma/E)xy + A(x) , \tag{3.21}$$

where $A(x)$ is an arbitrary function. Consequently,

$$u(x,y) = - \eta_0'(x)y - (1/E)p'(x)(h - y/2)y +$$
$$(\nu/E)P(x) + A(x) . \tag{3.22}$$

On the other hand, since the inelastic component

$$\epsilon^a_{xx} = \varphi(x,y) = u_{,x} - \epsilon^e_{xx} =$$

$$- [\nu\gamma/E + \eta_0''(x)]y - (1/E)p''(x)[h-y/2]y + A'(x) \tag{3.23}$$

must be non-negative, A(x) must satisfy the inequality

$$A'(x) \geq \psi(x,y) \, , \tag{3.24}$$

where

$$\psi(x,y) = [(\nu\gamma/E) + \eta_0''(x)]y + (1/E)p''(x)(h - y/2)y \, . \tag{3.25}$$

In agreement with (3.7), (3.9), (3.13) and (3.14), the third component w(x,y,z) of the displacement must satisfy the equations

$$w_{,z} = (\nu/E)[\gamma y + p(x)] + \zeta(x,y,z) \, , \tag{3.26}$$

where the inelastic strain $\epsilon^a{}_{zz}$ is denoted by $\zeta(x,y,z)$, and

$$w_{,x} = w_{,y} = 0 \, . \tag{3.27}$$

This implies that

$$\zeta(x,y,z) = G'(z) - (\nu/E)[\gamma y + p(x)] \, . \tag{3.28}$$

In (3.28) G'(z) is the derivative of an arbitrary function G(z) such that

$$G'(z) \geq G_0 \, , \tag{3.29}$$

with

$$G_0 = (\nu/E)\Big[\gamma h + \max_{x\epsilon[-b,b]} \{p(x)\}\Big] \, . \tag{3.30}$$

Consequently,

$$w(x,y,z) = G(z) \, . \tag{3.31}$$

Thus, the displacement fields defined by (3.19), (3.22) and (3.31), with A(x) and G(z) satisfying inequalities (3.24) and (3.29), respectively, solve the problem. If the external load is sufficiently smooth, as assumed, these inequalities are certainly satisfied, for example, by functions

$$A(x) = A_0 x, \tag{3.32}$$

where A_0 is the maximum of $\psi(x,y)$ in Ω, and

$$G(z) = G_0 z. \tag{3.33}$$

Because A(x) and G(z) are not unique, it may be of some interest to establish to which displacement field, if any, the smallest inelastic strains $\overline{\epsilon^a}_{xx}$ and $\overline{\epsilon^a}_{zz}$ correspond; that is such that if ϵ^a_{xx} and ϵ^a_{zz} are the inelastic strains corresponding to any other displacement which solves the problem, we have everywhere in Ω

$$\overline{\epsilon^a}_{xx}(x,y) - \epsilon^a_{xx}(x,y) \leq 0 \, , \tag{3.34}$$

$$\overline{\epsilon^a}_{zz}(x,y) - \epsilon^a_{zz}(x,y) \leq 0 \, . \tag{3.35}$$

For the sake of simplicity, we shall suppose that $p(x) = p_0$ is a constant function. In this case, it is enough to take

$$G(z) = (\nu/E)[p_0 + \gamma h]z \tag{3.36}$$

to get the minimum inelastic strain ϵ^a_{zz}, that is

$$\overline{\epsilon^a}_{zz}(x,y) = (\nu\gamma/E)(h - y) \, . \tag{3.37}$$

The displacement

$$w(x,y,z) = (\nu/E)[\gamma h + p_0]z \, , \tag{3.38}$$

null on the plane $z = 0$ and with the minimum absolute value, corresponds to this choice.

In its turn, A(x) must verify the inequality

$$A'(x) \geq [\nu\gamma/E + \eta_0''(x)]y \, , \tag{3.39}$$

with

$$\eta_0''(x) = (b^2/2D_0)(\gamma h + p_0)(x^2/b^2 - 1) \leq 0 \, . \tag{3.40}$$

Let us distinguish between two cases. If

$$\nu\gamma/E \geq (b^2/2D_0)(\gamma h + p_0) \, , \tag{3.41}$$

the right member of (3.39) is non-negative and it is easy to prove that

$$A_0(x) = [(\nu\gamma/E)x + \eta_0'(x)]h \, , \tag{3.42}$$

with

$$\eta_0'(x) = (b^2/6D_0)(\gamma h + p_0)(x^2/b^2 - 3)x \, , \tag{3.43}$$

is the choice to which corresponds the minimum inelastic strain

$$\bar{\epsilon}^a_{xx}(x,y) = [(\nu\gamma/E) + \eta_o''(x)](h-y) \,, \tag{3.44}$$

as well as the displacement along axis x,

$$u(x,y) = \eta_o'(x)(h-y) + (\nu/E)(\gamma h + p_o)x \,, \tag{3.45}$$

null on the symmetry plane $z = 0$ and with the minimum absolute value. The deformed configuration of the block is shown in Fig. 2.

In the case in which inequality (3.41) is not verified, the right member of (3.39) is non-negative for

$$|x| \geq x_0 = b\sqrt{1 - \frac{2\nu\gamma D_0}{b^2E(\gamma h + p_0)}} \,. \tag{3.46}$$

In this case it is still easy to verify that to the choice

$$A_1(x) = \begin{cases} [(\nu\gamma/E)(x-x_0) + \eta_0'(x) - \eta_0'(x_0)]h & \text{for } |x| \geq x_0 \,, \\ 0 & \text{for } |x| < x_0 \,, \end{cases} \tag{3.47}$$

corresponds the minimum value of the inelastic strain,

$$\bar{\epsilon}^a_{xx}(x,y) = \begin{cases} [(\nu\gamma/E) + \eta_o''(x)](h-y) & \text{for } |x| > x_o \,, \\ -[(\nu\gamma/E) + \eta_o''(x)]y & \text{for } |x| \leq x_o \,, \end{cases} \tag{3.48}$$

as well as displacement $u(x,y)$ with the following minimum absolute value:

$$u(x,y) = \begin{cases} \eta_0'(x)[h-y] + (\nu/E)[\gamma h + p_0]x - [(\nu\gamma/E)x_0 + \\ \qquad \eta_0'(x_0)]h \quad \text{for } |x| > x_0 \\ -\eta_0'(x)y + (\nu/E)p_0x \quad \text{for } |x| \leq x_0 \,. \end{cases} \tag{3.49}$$

Unlike $A_0'(x)$, $A_1'(x)$ is not differentiable for $x = x_0$. Therefore, in the case in which inequality (3.41) is not satisfied, no regular displacement field exists which solves the problem being examined and minimizes the inelastic strain.

The solution found requires a few brief remarks.

Firstly, we see that the stress field that solves the equilibrium problem is a plane stress one, independently of thickness 2c of the block.

Moreover, if we make the flexural rigidity D_0 of the plate approach infinity, we obtain from (3.19) and (3.45) the displacement components

$$u_\infty(x,y) = (\nu/E)(\gamma h + p_0)x \,, \tag{3.50}$$

$$v_\infty(x,y) = (1/E)p_0(h - y) + (\gamma/2E)(h^2 - y^2) \,, \tag{3.51}$$

which coincide with those already established by Di Pasquale for a plane rectangular element resting on a rigid base ['1].

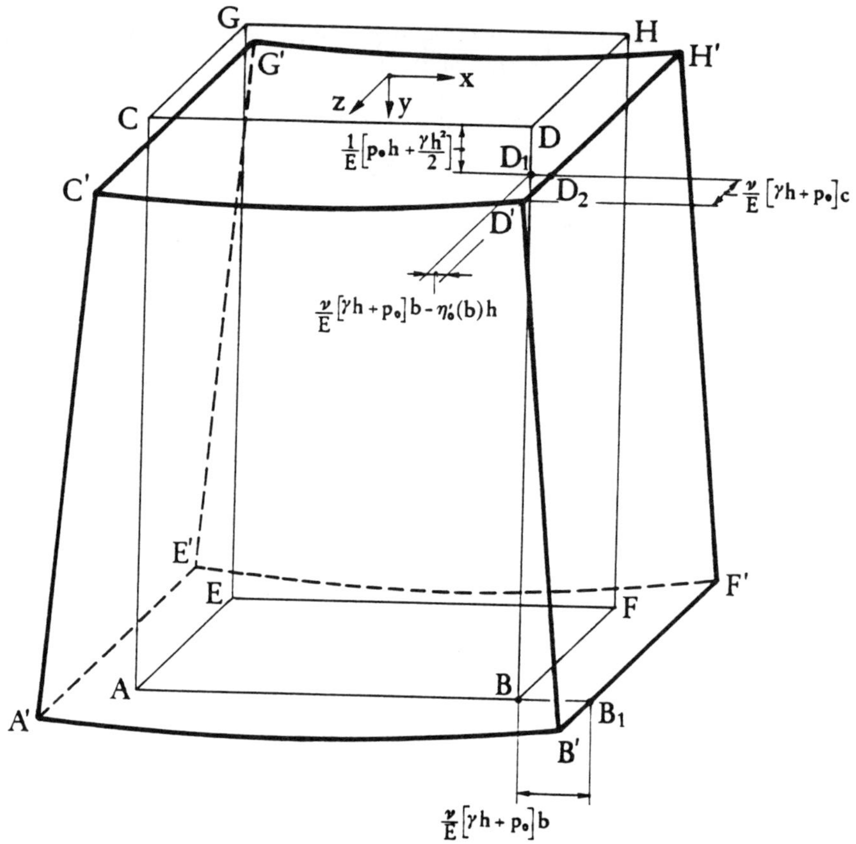

Fig. 2

Lastly, let us now show there exist necessary and sufficient conditions such that the anelastic deformation satisfies the compatibility equations. The following relations can be deduced from (3.23) and (3.28):

$$\epsilon^a_{xx,yy} = p''(x)/E \,,$$

$$\epsilon^a_{zz,yy} = \epsilon^a_{zz,xy} = 0 \,, \tag{3.52}$$

$$\epsilon^a_{zz,xx} = -\,\nu p''(x)/E \;;$$

therefore, the inelastic strains satisfy the compatibility equations only if we have

$$p''(x) = 0 , \tag{3.53}$$

or, considering (3.1), when p(x) is a constant function. Since the block is simply connected, the latter is a necessary and sufficient condition for the existence of a regular displacement field $\mathbf{u}^a(x,y,z)$ whose gradient has a symmetric part coinciding with $\mathbf{E}^a$.

4. The case of a wedge resting on an elastic shell

A wedge Λ of infinite thickness, whose transversal section is a sector AOB with opening 2β and radius R, is resting without friction on a cylindrical elastic shell that, in its turn, is resting on two supports situated below sides AC and BD (Fig. 3). The wedge, consisting of masonry-like material of negligible weight, is subjected to vertical force $f_0\mathbf{i}$, with $f_0 > 0$, distributed along edge OO'.

Since the thickness is infinite, the problem is a plane strain one. In the cylindrical reference system (r, θ, z) shown in Fig.3,

$$\mathbf{u}(r,\theta,z) = u(r,\theta)\mathbf{e}_r + v(r,\theta)\mathbf{e}_\theta , \tag{4.1}$$

where $u(r,\theta)$ and $v(r,\theta)$ are the components of the displacement $\mathbf{u}$ in the radial and tangential directions, indicated by $\mathbf{e}_r$ and $\mathbf{e}_\theta$. Consequently,

$$\epsilon_{zz} = \epsilon_{z\theta} = \epsilon_{zr} = 0 , \tag{4.2}$$

and, in agreement with the constitutive equation and with what has been proved in Section 2

$$\epsilon^a{}_{zz} = \epsilon^a{}_{z\theta} = \epsilon^a{}_{zr} = 0 . \tag{4.3}$$

On the other hand, using exactly the same line of reasoning as in the previous section, it is easy to verify that the only negative semi-definite stress field in equilibrium with the external loads is the following:

$$\sigma_{rr} = -(\xi f_0 \cos\theta)/r , \quad \sigma_{zz} = \nu\sigma_{rr} , \quad \sigma_{\theta\theta} = \tau_{rz} = \tau_{r\theta} = \tau_{z\theta} = 0 , \tag{4.4}$$

where $\xi = (\beta + \frac{1}{2}\sin 2\beta)^{-1}$. Correspondingly, we have the following elastic and inelastic strains:

$$\epsilon^e{}_{rr} = -(1/rE)(1 - \nu^2)\xi f_0 \cos\theta , \tag{4.5}$$

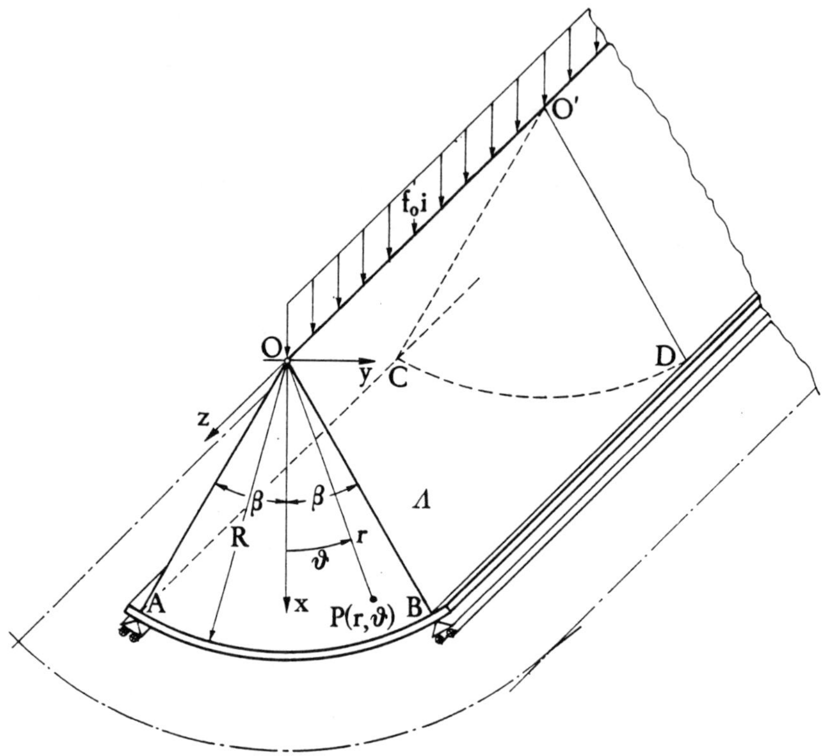

Fig.3

$$\epsilon^e{}_{\theta\theta} = (1/rE)\nu(1 + \nu)\xi f_0 \cos\theta , \tag{4.6}$$

$$\epsilon^e{}_{r\theta} = 0 , \tag{4.7}$$

$$\epsilon^a{}_{rr} = \epsilon^a{}_{r\theta} = 0 ; \qquad \epsilon^a{}_{\theta\theta} = \kappa(r,\theta) \geq 0 . \tag{4.8}$$

The components $u(r,\theta)$ and $v(r,\theta)$ of the displacement satisfy the differential equations

$$u_{,r} = \epsilon_{rr} = - (1/rE)(1 - \nu^2)\xi f_0 \cos\theta ,$$

$$u/r + v_{,\theta}/r = \epsilon_{\theta\theta} = (1/rE)\nu(1 + \nu)\xi f_0 \cos\theta + \kappa(r,\theta) , \tag{4.9}$$

$$u_{,\theta}/r + v_{,r} - v/r = 2\epsilon_{r\theta} = 0 .$$

Moreover,

$$u(R,\theta) = \psi(\theta) \tag{4.10}$$

must hold, where

$$\psi(\theta) = (f_0\xi R^3/D_1)[\cos\theta - \cos\beta + \tfrac{1}{2}\cos\beta(\theta^2 - \beta^2)] \tag{4.11}$$

is the radial displacement of the elastic shell with flexural rigidity D_1. Thus, integrating $(4.9)_1$ and bearing in mind (4.10), we deduce

$$u(r,\theta) = (1/E)(1 - \nu^2)\xi f_0\cos\theta\ln(R/r) + \psi(\theta) . \tag{4.12}$$

Then, integrating $(4.9)_2$ we obtain

$$v(r,\theta) = (1/E)\nu(1 + \nu)\xi f_0\sin\theta + rK(r,\theta) +$$

$$- (1/E)(1 - \nu^2)\xi f_0\sin\theta\ln(R/r) - \Psi(\theta) + B(r) , \tag{4.13}$$

where $K(r,\theta)$ is a primitive of $\kappa(r,\theta)$ with respect to θ, $\Psi(\theta)$ is a primitive of $\psi(\theta)$ and $B(r)$ is an arbitrary function.

On the other hand, $(4.9)_3$, (4.12) and (4.13) give the equation

$$r^2K(r,\theta)_{,r} + \Psi(\theta) + \Psi''(\theta) + (1/E)(1 + \nu)(1 - 2\nu)\xi f_0\sin\theta +$$

$$- B(r) + rB'(r) = 0 , \tag{4.14}$$

which, in its turn, implies:

$$K(r,\theta) = (f_0\xi/r)[(1/E)(1 + \nu)(1 - 2\nu)\sin\theta +$$

$$(R^3/2D_1)\cos\beta(\tfrac{1}{3}\theta^2 - \beta^2)\theta] - B(r)/r + \chi(\theta) , \tag{4.15}$$

where χ is an arbitrary function of θ. Consequently,

$$v(r,\theta) = (1/E)(1 - \nu^2)\xi f_0\sin\theta[1 - \ln(R/r)] +$$

$$- \xi f_0(R^3/D_1)(\sin\theta - \theta\cos\beta) + r\chi(\theta) . \tag{4.16}$$

On the other hand, since the inelastic strain $\epsilon^a{}_{\theta\theta}$ must be non-negative, $\chi(\theta)$ must satisfy the inequality

$$\chi'(\theta) + (f_0\xi/r)\omega(\theta) \geq 0 , \tag{4.17}$$

with

$$\omega(\theta) = (1/E)(1 + \nu)(1 - 2\nu)\cos\theta + (R^3/2D_1)\cos\beta(\theta^2 - \beta^2) . \tag{4.18}$$

Condition (4.17) can be verified everywhere in Λ only if $\omega(\theta) \geq 0$ for each $\theta \in [-\beta, \beta]$. If this condition is not satisfied, since $\omega(\theta)$ is a bounded function, for each $\delta > 0$ it is possible to choose $\chi'(\theta)$ so that condition (4.17) is satisfied at every point of Λ with $r \geq \delta$.

In this case, too, it may be useful to establish which of all the functions $\chi'(\theta)$ that satisfy condition (4.17), is the one which the minimum inelastic strain corresponds to. For the sake of simplicity, we shall confine ourselves to the case in which

$$\beta \leq \sqrt{2}\,. \tag{4.19}$$

We must distinguish between two cases. If

$$D_1 \geq \frac{R^3 E \beta^2 \cos\beta}{2(1+\nu)(1-2\nu)}\,, \tag{4.20}$$

bearing (4.19) in mind, a direct calculation shows that $\omega(\theta)$ is non-negative; therefore, the minimum inelastic strain corresponds to the function

$$\chi_0'(\theta) = -\, f_0 \xi \omega(\theta)/R\,, \tag{4.21}$$

satisfying condition (4.17). If, on the contrary, we have

$$D_1 < \frac{R^3 E \beta^2 \cos\beta}{2(1+\nu)(1-2\nu)}\,, \tag{4.22}$$

it can be deduced from (4.18) and (4.19) that

$$\omega(0) < 0\,;\quad \omega(\beta) > 0\,;\quad \omega'(\theta) \geq 0 \quad \text{for each } \theta \in [0, \beta]\,, \tag{4.23}$$

so that a unique $\theta_0 \in (0, \beta)$ exists, such that

$$\omega(\theta) \begin{cases} \leq 0 & \text{for } \theta \in [-\theta_0, \theta_0]\,, \\ > 0 & \text{for } \theta \in [-\beta, -\theta_0) \text{ and } \theta \in (\theta_0, \beta]\,. \end{cases} \tag{4.24}$$

Thus, if $\delta > 0$ is given, the function

$$\chi_1'(\theta) = \begin{cases} -\, f_0 \xi \omega(\theta)/\delta & \text{for } \theta \in [-\theta_0, \theta_0]\,, \\ -\, f_0 \xi \omega(\theta)/R & \text{for } \theta \in [-\beta, -\theta_0) \text{ and } \theta \in (\theta_0, \beta]\,, \end{cases} \tag{4.25}$$

continuous everywhere but not differentiable for $\theta = \pm\,\theta_0$, satisfying inequality (4.17) in every point of Λ with $r \geq \delta$, minimizes the inelastic strain $\epsilon^a_{\theta\theta}$.

Since function $\chi(\theta)$ is known, the component $v(r,\theta)$ of the displacement can be calculated from (4.16).

In particular, when flexural rigidity D_1 approaches infinity, we obtain the displacements

$$u_\infty(r,\theta) = (1/E)(1-\nu^2)\xi f_0 \cos\theta \ln(R/r)\ , \qquad (4.26)$$

$$v_\infty(r,\theta) = v_\infty(r,0) + (1/E)\xi f_0 \sin\theta\{(1-\nu^2)[1-\ln(R/r)] + $$
$$- (1+\nu)(1-2\nu)(r/R)\}\ , \qquad (4.27)$$

corresponding to the case in which the wedge is resting on a rigid cylindrical support.

The relation

$$\epsilon^a_{\theta\theta,rr} + (2/r)\epsilon^a_{\theta\theta,r} = 0 \qquad (4.28)$$

can easily be deduced from $(4.8)_3$ and (4.15). Therefore the inelastic strains satisfy the compatibility equations, as a direct calculation shows. Since Λ is simply connected, for each regular solution there exist two functions u^a and v^a such that, for $r > 0$, we have

$$u^a_{,r} = 0\ ,$$

$$u^a/r + v^a_{,\theta}/r = \epsilon^a_{\theta\theta}\ , \qquad (4.29)$$

$$u^a_{,\theta}/r + v^a_{,r} - v^a/r = 0\ .$$

References

[1] Romano G., Romano M. *Sulla soluzione di problemi strutturali in presenza di legami costitutivi unilaterali.* Rend. Acc. Naz. dei Lincei, Classe Scienze Fisiche, Matematiche e Naturali, Serie VIII, vol. LXVII, 1979.

[2] Di Pasquale S. *Questioni di meccanica dei solidi non reagenti a trazione.* Atti del VII Congresso Nazionale AIMETA, vol. I, pp. 251-263, Genova 1982.

[3] Abruzzese D., Grimaldi A., Sacco E. *Indagine numerica su alcuni problemi di materiali non resistenti a trazione.* Int. Rep. 14, Dip. di Ingegneria Civile Edile, II Universita' di Roma, 1987.

[4] Del Piero G. *Constitutive equation and compatibility of the external loads for linear elastic masonry-like materials.* Meccanica, **24**, pp. 150-162, 1989.

[5] Giaquinta M., Giusti E., Researches on the equilibrium of masonry structures. Arch. Rat. Mech. Anal. **68**, pp. 359-392, 1985.

[6] Anzellotti G., *A class of non-coercive functionals and masonry-*

like materials. Ann. Inst. H. Poincare' **2**, pp.261-307, 1985.
[7] Del Piero G. *Recent developments in the mechanics of materials which do not support tension*. Proc. Int. Colloq. on free boundary problems. IRSEE, June 1987.
[8] Villaggio P. *Stress diffusion in masonry walls*. J. Struct. Mech., **9**, pp. 439-450, 1981.
[9] Bennati S., Lucchesi M., *The minimal section of a triangular masonry dam*. Meccanica, **23**, pp. 221-225, 1988.
[10] Bennati S., Gennai A. M., Padovani C. *Trapezoidal gravity dams working in a state of pure compression*. Mech. Struct. and Mach. To appear.
[11] Di Pasquale S. *Statica dei solidi murari: teoria ed esperienze*. Int. Rep. Dip. diCostruzioni Universita' di Firenze, 1984.
[12] Gennai A. M., Padovani C., *Soluzioni esplicite di equazioni costitutive per materiali non reagenti a trazione*. Atti del IX Congresso Nazionale AIMETA, vol. I, pp. 129-132, Bari 1988.

Istituto di Scienza delle Costruzioni
via Diotisalvi 2, 56100-PISA

CNUCE- C.N.R.
via S. Maria 36, 56100-PISA

International Series of Numerical Mathematics, Vol. 101, © 1991 Birkhäuser Verlag Basel

FINITE ELEMENT ANALYSIS OF NONLINEAR LAMINATED COMPOSITE PLATES

by

D. BRUNO, G. PORCO and R. ZINNO

Department of Structures – University of Calabria
Arcavacata di Rende (Cosenza) – Italy

Summary. In this paper a finite element analysis of cross-ply laminated bimodulus composite plates under transverse loads is presented.

The analysis takes into account the transverse shear (as in the Mindlin-Reissner plate theory) and large displacements (in the sense of the von Karman theory) suitable for simulating the behaviour of moderately thick plates.

Numerical results are given relative to various boundary conditions, together with some comparisons with previous results available in literature.

Sommario. In questo lavoro viene sviluppata una analisi agli elementi finiti di piastre laminate composite bimodulari del tipo "cross-ply" soggette a carichi trasversali.

L' analisi e' sviluppata nell' ambito della teoria delle piastre moderatamente spesse, tenendo conto delle deformazioni dovute al taglio (nel senso della teoria delle piastre spesse di Mindlin-Reissner) e di grandi spostamenti (nel senso della teoria di von Karman).

Vengono forniti alcuni risultati numerici per varie condizioni di vincolo, nonchè alcuni confronti con precedenti risultati disponibili in letteratura.

NOTATION

A_{ij}, B_{ij}, D_{ij} = Extensional, flexural-extensional and flexural stiffnesses (i, j = 1, 2, 3).

H_{ij} = Shear stiffness (i, j = 4, 5).

a, b = Plate dimensions along the x and y directions, respectively.

E_{1C}, E_{1T} = Layer elastic moduli in directions along fibers in compression and in tension.

E_{2C}, E_{2T} = Layer elastic moduli in directions normal to fibers in compression and in tension.

G_{12a}, G_{13a}, G_{23a} = Layer in-plane and thickness shear moduli (a = C in compression, a = T in tension).

h = Total thickness of the laminate.

t_k	= Thickness of $k-th$ layer.
K	= Shear correction coefficient.
x, y, z	= Position coordinates in a cartesian system.
M_i, N_i	= Stress couples and stress resultants, respectively ($i = x, y, xy$).
Q_i	= Shear stress resultants ($i = x, y$).
u, v, w	= Displacements in the x, y, z directions, respectively.
u_o, v_o, w_o	= Displacements of the midplane in the x, y, z directions, respectively.
θ^k	= Orientation of the $k-th$ layer.
ψ_x, ψ_y	= Slopes in the (x,z) and (y,z) planes.
χ_x, χ_y, χ_{xy}	= Curvature components.
ε_x, ε_y, ε_{xy}	= Strain components.
ε_x^o, ε_y^o, ε_{xy}^o	= Middle surface strain components.
n	= Number of layers.
q_o	= Uniform transverse load.
N	= Number of global nodes of the mesh in the *F.E.M.* discretization.
ν_{12C}, ν_{12T}	= Major Poisson's ratios for orthotropic bimodulus material.
$\overline{C}_{ijl}^k$	= Material stiffness coefficients (local system) of the $k-th$ layer ($l = 1$ in compression, $l = 2$ in tension) .
C_{ijl}^k	= Material stiffness coefficients (global system) of the $k-th$ layer ($l = 1$ in compression, $l = 2$ in tension).
T^k	= Transformation matrix of the $k-th$ layer.
f_i	= Interpolation functions.
z_{nx}, z_{ny}	= Neutral surface positions associated with $\varepsilon_x = 0$ and $\varepsilon_y = 0$, respectively.

1. Introduction

Laminated composite plates are receiving increasing attention in structural applications because of their high stiffness-to-weight ratio and the capability of the anisotropic properties to be tailored through variation of the fibre orientation and stacking sequence.

The shear effects are more pronounced in composite laminated plates than in isotropic plates because of their low transverse shear moduli relative to the in-plane Young's moduli. The classical thin plate theory based on the Kirchhoff-Love assumption that normals to the middle surface before deformation remain straight and normal to it after deformation, is not adequate for the flexural analysis of moderately thick laminates [1 ÷ 4].

Certain fibre-reinforced composite materials exhibit quite different elastic behaviour in tension and compression [5-6]. This behaviour is usually modeled with a bilinear stress-strain relationship with moduli E_c in compression and E_t in tension (Fig. 1). This model was found to agree with experimental results. Such materials are called "Bimodulus composite materials" and their analysis is more difficult than ordinary materials since the elastic moduli depend on the sign of the stresses which are unknown *"a priori"*.

The aim of the present work is the analysis of the behaviour of bimodular composite plates arbitrarily laminated in a cross-ply fashion and subject to transverse loads. The analysis is developed using a finite element procedure based on a theory taking account of transverse shear deformability and including nonlinear terms in the sense of von Karman

in the strain- displacement relationship.
Numerical results relative to rectangular plates with various edge conditions are presented.

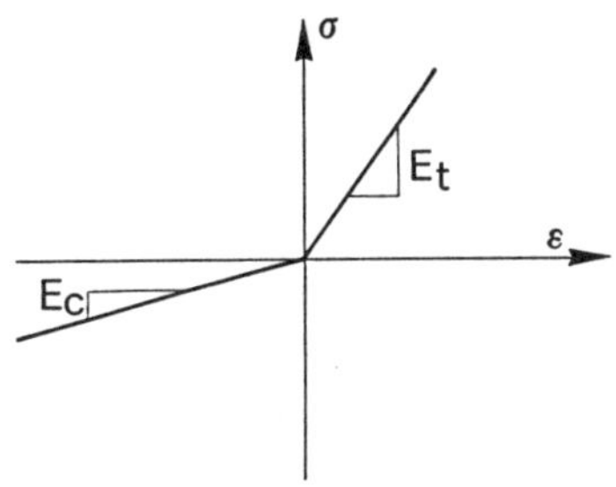

Fig. 1 - Stress-strain curve for bimodular material

2. Formulation

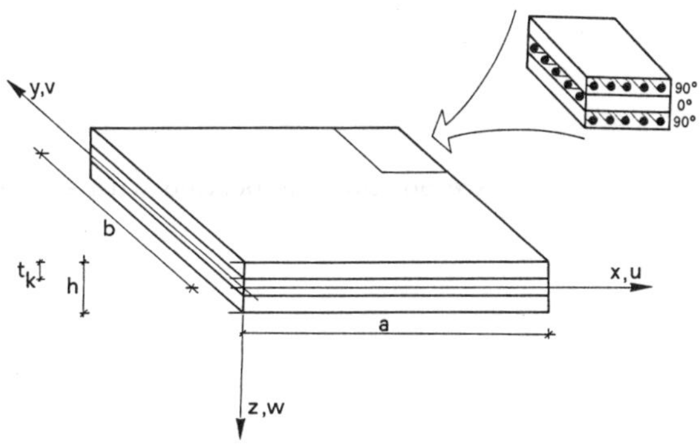

Fig. 2 - Plate model

With reference to Fig. 2 a laminated "cross-ply" plate is examined.
The plate under consideration consists of n layers alternately oriented at 0° and 90° to each one of the plate edges. Each layer has different orthotropic moduli in tension and compression. The coordinate system (x, y, z) is taken as in Fig. 2. In particular, the plain $z = 0$ is coincident with the middle plane of the plate. Let Ω be the middle surface. Many results relative to the behaviour of laminates made of materials with the same moduli in tension and compression [7-8-9] are known.

For symmetric laminates, there is no coupling between bending and extension. For unsymmetric laminates the effect on the deflections of coupling between bending and extension decays very rapidly as the number of layers increases. The same conclusions are not valid for laminates made of materials with different moduli in tension and compression. For example, a laminate which is symmetric in geometry and in material properties with respect to the middle surface, exibits coupled beetwen bending and stretching when

subject to transverse loads.
The displacement field is assumed to be of the form:

$$\begin{aligned} u(x,y,z) &= u_o(x,y) + z\psi_x(x,y), \\ v(x,y,z) &= v_o(x,y) + z\psi_y(x,y), \\ w(x,y,z) &= w_o(x,y), \end{aligned} \tag{2.1}$$

where u, v, w are, respectively, the displacements in the x, y, z directions, u_o, v_o, w_o are the corresponding midplane displacements and ψ_x, ψ_y are the bending slopes in the (x,z) and (y,z) planes.

Assuming that the transverse deflection is comparable to the total thickness of the plate and that strains are much smaller than rotations, the nonlinear strain-measure can be taken as:

$$\begin{aligned} \varepsilon_{xx} &= u_{o,x} + \tfrac{1}{2}w_{o,x}^2 + z\psi_{x,x} = \varepsilon_x^o + z\chi_x , \\ \varepsilon_{yy} &= v_{o,y} + \tfrac{1}{2}w_{o,y}^2 + z\psi_{y,y} = \varepsilon_y^o + z\chi_y , \\ \varepsilon_{zz} &= \tfrac{1}{2}\left(\psi_x^2 + \psi_y^2\right), \\ 2\varepsilon_{xz} &= \psi_x + w_{,x} = \gamma_{xz} , \\ 2\varepsilon_{yz} &= \psi_y + w_{,y} = \gamma_{yz} , \\ 2\varepsilon_{xy} &= u_{o,y} + v_{o,x} + w_{o,x}w_{o,y} + z(\psi_{x,y} + \psi_{y,x}) = \varepsilon_{xy}^o + z\chi_{xy} . \end{aligned} \tag{2.2}$$

Furthermore, ε_{zz} is neglected since the constitutive relations are based on the plane stress assumption. With reference to Fig. 3, the constitutive equations of the $k-th$ layer are:

$$\begin{Bmatrix} \overline{\sigma}_{11l}^k \\ \overline{\sigma}_{22l}^k \\ \overline{\sigma}_{12l}^k \\ \overline{\sigma}_{13l}^k \\ \overline{\sigma}_{23l}^k \end{Bmatrix} = \begin{bmatrix} \overline{C}_{11l}^k & \overline{C}_{12l}^k & 0 & 0 & 0 \\ \overline{C}_{21l}^k & \overline{C}_{22l}^k & 0 & 0 & 0 \\ 0 & 0 & \overline{C}_{33l}^k & 0 & 0 \\ 0 & 0 & 0 & \overline{C}_{44l}^k & 0 \\ 0 & 0 & 0 & 0 & \overline{C}_{55l}^k \end{bmatrix} \begin{Bmatrix} \overline{\varepsilon}_{11l}^k \\ \overline{\varepsilon}_{22l}^k \\ \overline{\varepsilon}_{12l}^k \\ \overline{\varepsilon}_{13l}^k \\ \overline{\varepsilon}_{23l}^k \end{Bmatrix}, \tag{2.3}$$

where $\overline{\sigma}_{ijl}^k$ and $\overline{\varepsilon}_{ijl}^k$ are the components of the stress and strain tensors, respectively, defined in the material coordinates, and $\overline{C}_{ijl}^k$ are the material stiffness coefficients.

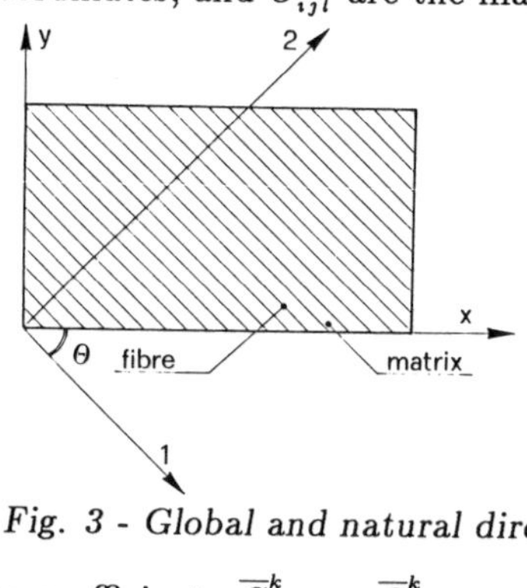

Fig. 3 - Global and natural directions

For the $k-th$ layer, the coefficients $\overline{C}_{ijl}^k = \overline{C}_{jil}^k$ are given (in terms of engineering constants) by:

$$
\begin{aligned}
&\overline{C}^k_{11l} = E^k_{1l}/(1-\nu^k_{12l}\nu^k_{21l}), \qquad \overline{C}^k_{12l} = \nu^k_{12l}E^k_{2l}/(1-\nu^k_{12l}\nu^k_{21l}),\\
&\overline{C}^k_{21l} = \nu^k_{21l}E^k_{1l}/(1-\nu^k_{12l}\nu^k_{21l}), \qquad \overline{C}^k_{22l} = E^k_{2l}/(1-\nu^k_{12l}\nu^k_{21l}),\\
&\overline{C}^k_{33l} = G^k_{12l}, \qquad \overline{C}^k_{44l} = G^k_{13l}, \qquad \overline{C}^k_{55l} = G^k_{23l},
\end{aligned}
\tag{2.4}
$$

where $G^k_{12l}, G^k_{13l}, G^k_{23l}$ are in-plane and thickness shear moduli and E^k_{1l}, E^k_{2l} are the elastic moduli in the directions parallel and normal to the fibres, respectively. The coefficient l can assume two values $l = 1$ in compression, $l = 2$ in tension.

For a unimodular orthotropic elastic medium (same moduli in tension and compression), in the case of constitutive relations based on the plane-stress assumption there are only six independent elastic constants. For a bimodular orthotropic elastic medium there are twelve independent elastic constants, six in compression and six in tension.
For the $k-th$ layer, the stress-strain relation is:

$$
\{\sigma^k\} = [T^k]\left[\overline{C}^k\right][T^k]^{-1}\{\varepsilon^k\} = [C^k]\{\varepsilon^k\}, \tag{2.5}
$$

where $\{\sigma^k\}$ and $\{\varepsilon^k\}$ are the components of stress and strain tensors defined with reference to the plate axes:

$$
\{\sigma^k\} = \begin{Bmatrix} \sigma^k_x \\ \sigma^k_y \\ \tau^k_{xy} \\ \tau^k_{xz} \\ \tau^k_{yz} \end{Bmatrix}, \qquad \{\varepsilon^k\} = \begin{Bmatrix} \varepsilon^k_x \\ \varepsilon^k_y \\ 2\varepsilon^k_{xy} \\ 2\varepsilon^k_{xz} \\ 2\varepsilon^k_{yz} \end{Bmatrix}, \tag{2.6}
$$

and the transformation matrix T^k is given by:

$$
[T^k] = \begin{bmatrix} F^k & P^k & 2R^k & 0 & 0 \\ P^k & F^k & -2R^k & 0 & 0 \\ -R^k & R^k & S^k & 0 & 0 \\ 0 & 0 & 0 & m^k & n^k \\ 0 & 0 & 0 & -n^k & m^k \end{bmatrix}, \tag{2.7}
$$

with:

$$
\begin{aligned}
&m^k = cos\theta^k, \qquad n^k = sen\theta^k, \qquad F^k = cos^2\theta^k,\\
&P^k = sen^2\theta^k, \qquad R^k = sen\theta^k cos\theta^k, \qquad S^k = cos^2\theta^k - sen^2\theta^k.
\end{aligned}
\tag{2.8}
$$

The explicit form of the coefficients C^k_{ij} are given in Ref [9].

The total potential energy of the plate, in absence of body forces, is given by:

$$
\begin{aligned}
\Pi = \frac{1}{2}\int_\Omega &(N_x\varepsilon^o_x + N_y\varepsilon^o_y + N_{xy}\varepsilon^o_{xy} + M_x\chi_x + M_y\chi_y + M_{xy}\chi_{xy} +\\
&+Q_x\gamma_{xz} + Q_y\gamma_{yz})d\Omega - \int_\Omega q_o w d\Omega - \int_{\Gamma_1}\overline{N}_n u_n ds - \int_{\Gamma_2}\overline{N}_s u_s ds +\\
&-\int_{\Gamma_3}\overline{M}_n\psi_n ds - \int_{\Gamma_4}\overline{M}_s\psi_s ds - \int_{\Gamma_5}\overline{Q}_n w ds,
\end{aligned}
\tag{2.9}
$$

where Γ_i are the portions of the boundary Γ of the middle surface Ω on which $\overline{N}_n$,

$\overline{N}_s, \overline{M}_n, \overline{M}_s, \overline{Q}_n$, respectively, are specified. Furthermore, q_o is the normal pressure and

$$(N_i, M_i) = \int_{-h/2}^{h/2} (1, z)\sigma_i dz \qquad (i = x, y, xy), \tag{2.10}$$

$$(Q_x, Q_y) = \int_{-h/2}^{h/2} (\tau_{zx}, \tau_{zy}) dz. \tag{2.11}$$

From eqs. (2.5), (2.10), (2.11) we obtain the plate constitutive equations:

$$\left\{ \begin{matrix} N_i \\ M_i \end{matrix} \right\} = \begin{bmatrix} A_{ij} & B_{ij} \\ B_{ij} & D_{ij} \end{bmatrix} \left\{ \begin{matrix} \varepsilon_j^o \\ \chi_j \end{matrix} \right\}, \tag{2.12}$$

$$\left\{ \begin{matrix} Q_x \\ Q_y \end{matrix} \right\} = \begin{bmatrix} H_{44} & H_{45} \\ H_{54} & H_{55} \end{bmatrix} \left\{ \begin{matrix} \gamma_{xz} \\ \gamma_{yz} \end{matrix} \right\}. \tag{2.13}$$

Here $A_{ij}, B_{ij}, D_{ij}\,(i,j = 1,2,3)$ and $H_{ij}\,(i,j = 4,5)$ are the extensional, flexural - extensional, flexural and shear stiffnesses respectively :

$$A_{ij} = \sum_{k=1}^{n} \int_{h_{k-1}}^{h_k} C_{ijl}^k dz, \qquad B_{ij} = \sum_{k=1}^{n} \int_{h_{k-1}}^{h_k} C_{ijl}^k z dz, \tag{2.14 a, b}$$

$$D_{ij} = \sum_{k=1}^{n} \int_{h_{k-1}}^{h_k} C_{ijl}^k z^2 dz, \qquad H_{ij} = K^2 \sum_{k=1}^{n} \int_{h_{k-1}}^{h_k} C_{ijl}^k dz. \tag{2.14 c, d}$$

$K^2 = 5/6$ is the square of the shear correction coefficient. The symbols h_k and h_{k-1} denote the distances from the midplane to the lower and upper surface of the k-th layer, respectively.

In laminates made of bimodular materials, the plate stiffnesses are not known a priori. Indeed, the stiffnesses of each layer depends upon whether it is located above or below the neutral surface.

The method of evaluating the coefficients A_{ij}, B_{ij}, D_{ij} and H_{ij} is based on the fibre-governed model proposed by Bert in which the evalutation of integrals is made only in fiber directions for each layer [6]. This model has been shown to agree well with experimental results for several materials with drastically different elastic properties in tension and in compression. With reference to Fig. 4 the coefficients A_{ij} can be written as:

$$A_{ij} = \int_{h_1}^{z_{ny}} C_{ij1}^1 dz + \int_{z_{ny}}^{h_2} C_{ij2}^1 dz + \int_{h_2}^{h_3} C_{ij1}^2 dz + \int_{h_3}^{h_4} C_{ij2}^3 dz \qquad (i,j = 1,2,3). \tag{2.15}$$

In a similar way the next three integrals in equations (2.14) can be evaluated. Here z_{nx} and z_{ny} denote the neutral-surface positions associated with $\varepsilon_x = 0$ and $\varepsilon_y = 0$, respectively.

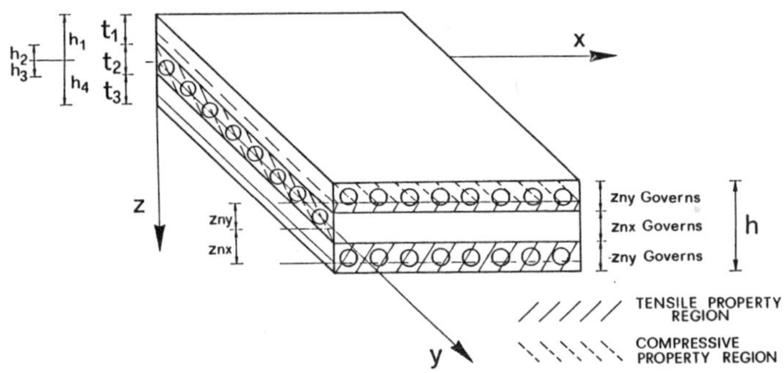

Fig. 4 - Scheme of an element of a three - layer cross-ply plate.

A variational formulation of the plate equilibrium equations is obtained by applying the stationarity condition to the total potential energy functional (2.9):

$$\delta\Pi = \int_{\Omega} (N_x \delta\varepsilon_x^o + N_y \delta\varepsilon_y^o + N_{xy} \delta\varepsilon_{xy}^o + M_x \delta\chi_x + M_y \delta\chi_y + M_{xy} \delta\chi_{xy} + \\ + Q_x \delta\gamma_{xz} + Q_y \delta\gamma_{yz} - q_o \delta w) d\Omega - \int_{C_1} \overline{N}_n \delta u_n \, ds - \int_{C_2} \overline{N}_s \delta u_s \, ds + \\ - \int_{C_3} \overline{M}_n \delta\psi_n \, ds - \int_{C_4} \overline{M}_s \delta\psi_s \, ds - \int_{C5} \overline{Q}_n \delta w ds = 0. \tag{2.16}$$

3. Finite element formulation

In this section a finite element model based on the variational formulation of the equation (2.16) is presented.

The region Ω is divided into a finite number of isoparametric rectangular elements. Over each element the generalized displacements $(u, v, w, \psi_x, \psi_y)$ are interpolated as:

$$u = \sum_{i=1}^{N} u_i f_i , \qquad v = \sum_{i=1}^{N} v_i f_i , \qquad w = \sum_{i=1}^{N} w_i f_i ,$$

$$\psi_x = \sum_{i=1}^{N} \psi_{xi} f_i , \qquad \psi_y = \sum_{i=1}^{N} \psi_{yi} f_i , \tag{3.1}$$

where $u_i, v_i, w_i, \psi_{xi}, \psi_{yi}$ are the values of the unknown functions at the nodes of the mesh and f_i are the interpolation functions.

The substitution of the relations (3.1) into Eq. (2.16) gives:

$$\mathbf{K} \ \mathbf{U} = \mathbf{F} , \tag{3.2}$$

where $\mathbf{U}$ collects the nodal values of the generalized displacements u, v, w, ψ_x, ψ_y, $\mathbf{K}$ is the stiffness matrix of the plate, and $\mathbf{F}$ is the nodal force vector.

Notice that the stiffness matrix **K** depends on the solution **U**. Therefore, a standard iterative procedure must be used.

4. Numerical results.

The following material properties typical of advanced bimodulus fibre-rinforced composites are assumed.

Tab. I - Material properties of laminae

PROPERTY and UNITS	ARAMID - RUBBER		POLYESTER - RUBBER	
	TENSION	COMPRESSION	TENSION	COMPRESSION
E_1 - GPa	3.5842	0.0120	0.617	0.0369
E_2 - GPa	0.00909	0.0120	0.00800	0.0106
G_{12}-GPa	0.00370	0.00370	0.00262	0.00267
G_{13}-GPa	0.00370	0.00370	0.00262	0.00267
G_{23}-GPa	0.00290	0.00499	0.00233	0.00475
ν_{12}	0.416	0.205	0.475	0.185

Here we give some numerical results for simply supported (SS) and clamped (CC) rectangular plates under uniform transverse load. The constraints are defined by the following schemes.

Tab. II - Boundary conditions

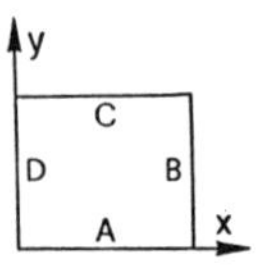

	SIDE A	SIDE B	SIDE C	SIDE D
SS	$u,w,\psi_x=0$	$v,w,\psi_y=0$	$u,w,\psi_x=0$	$v,w,\psi_y=0$
CC	all edges clamped $u,v,w,\psi_x,\psi_y=0$			

At first, we give some results on the influence of the type of integration (reduced or full of the shear terms), and of the structure of the F.E.M. mesh.

In particular, the central deflection w_c of a simply supported two layer Aramid - Rubber square plate under uniform transverse load q_o is shown in tab III, in the nondimensional form $\overline{w}_c = w_c(E_{2C}h^3/q_o a^4)$ for various F.E.M. meshes (linear elements (L), quadratic elements (Q)) and for full and reduced integration of shear terms. It comes out that the influence of the integration of shear terms is more important for linear elements. As far as the F.E.M. mesh convergence is concerned it can be observed that 4x4Q and 8x8L give good accuracy in calculations. However, a 12x12L F.E.M. mesh and reduced integration for shear terms were employed in numerical applications.

Tab III - Influence of integration of shear terms and F.E.M. mesh on the central deflection $\overline{w}_c$

	a/h=10		a/h=100	
mesh	reduced	full	reduced	full
4x4 L	3.0215 E-2	1.9482 E-2	2.7129 E-2	4.6732 E-4
2x2 Q	3.1733 E-2	2.9684 E-2	2.8939 E-2	2.3916 E-2
8x8 L	3.0619 E-2	2.6898 E-2	2.7818 E-2	1.7800 E-3
4x4 Q	3.0801 E-2	3.0692 E-2	2.8071 E-2	2.7301 E-2
12x12 L	3.0694 E-2	2.8916 E-2	2.7900 E-2	3.7109 E-3
6x6 Q	3.0761 E-2	3.0742 E-2	2.8037 E-2	2.7743 E-2

In Figs. 5, 6, 7 some results for cross-ply rectangular plates subjected to a uniform transverse load q_o are shown. In particular, Fig. 5 shows the influence of the aspect ratio b/a; this is more important in the simply-supported plate; on the contrary, the influence of shear deformability is more important in the clamped plate. A comparison of results obtained in Ref. [12] with ones made in the present analysis is given in tables shown in the same figures. It can be observed that a good agreement between two results is found. The influence of the b/a aspect ratio and the boundary conditions on the nondimensional central deflections is similar for Aramid-Rubber and Polyester-Rubber plate's.

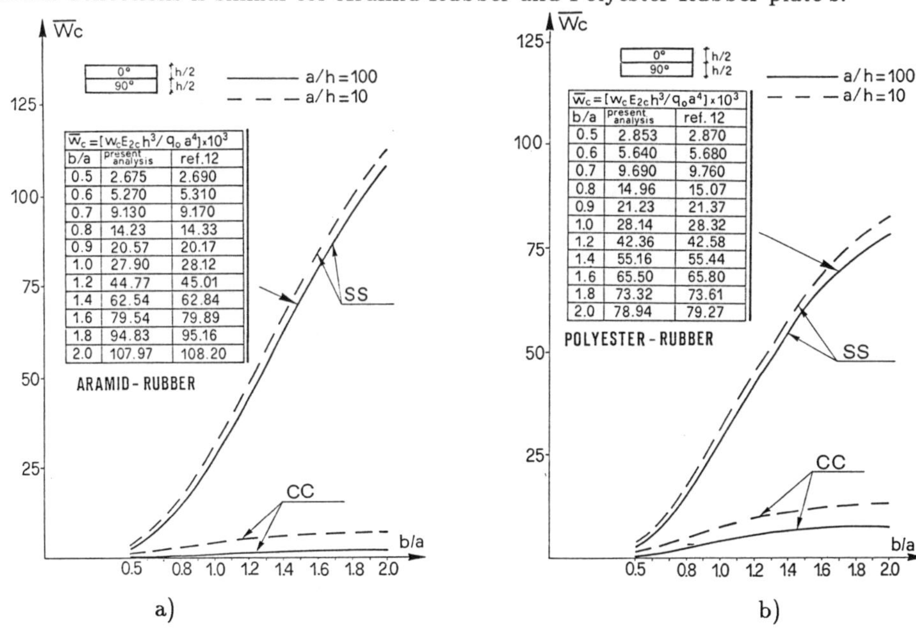

Fig. 5a, b - Cross-ply rectangular plate: influence of aspect ratio b/a on the central deflection $\overline{w}_c$

Fig. 6 refers to a cross-ply two-layer square plate. The symbol E_c (E_t) denote the plates made in unimodular material with only the six independent elastic constants in compression (in tension) of the bimodular material. In this figure the influence of the thickness ratio t_1/t_2 on the deformability is shown: the deformability is strongly influenced by t_1/t_2 in the case of bimodular plates, whereas it is almost independent of t_1/t_2 in the case of unimodular laminate.

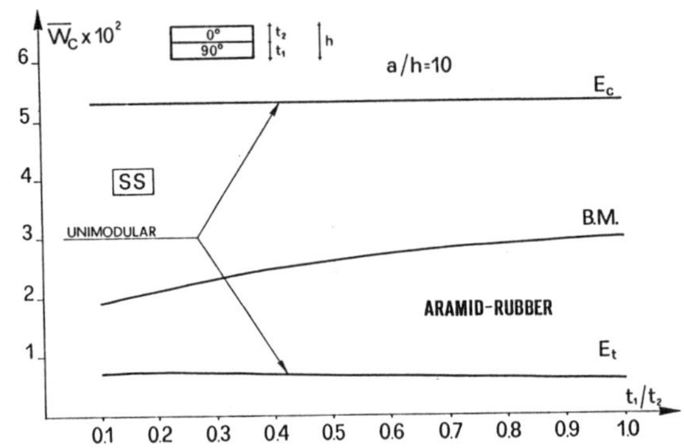

Fig. 6 - Influence of the thickness ratio t_1/t_2 on the deflection w_c

Tab. IV shows the influence of layering on deformability of a cross-ply square plate. A different behaviour can be observed between ordinary composite material plates and bimodular ones. This is due to the different bending-stretching coupling in the bimodular composites.

Tab. IV - Influence of layering on deformability.
(Aramid-Rubber, SS, $a/b = 1$, $a/h = 100$)

	$\overline{W}_C$		
	E_c	E_t	B. M.
2 layers	5.20112E-2	2.53235E-3	2.79487E-2
4 layers	5.20112E-2	8.34594E-4	2.16468E-2
25 layers	5.20112E-2	6.87535E-4	1.62412E-2

In Figs. 7a, b the influence of material and geometrical nonlinearities is examined. In particular, the load-deflection curves of an Aramid-Rubber cross-ply square plate are shown. Fig. 7a also shows the case of unimodular composite material plate (Modulus E_c). It can be observed that the influence of geometric nonlinearities is more significant in the case of unimodular materials.

Finally, in fig. 7b the influence of boundary conditions on geometrical nonlinear effects is shown. It was found that the influence of geometrical nonlinear effects is more important in the case of SS boundary conditions.

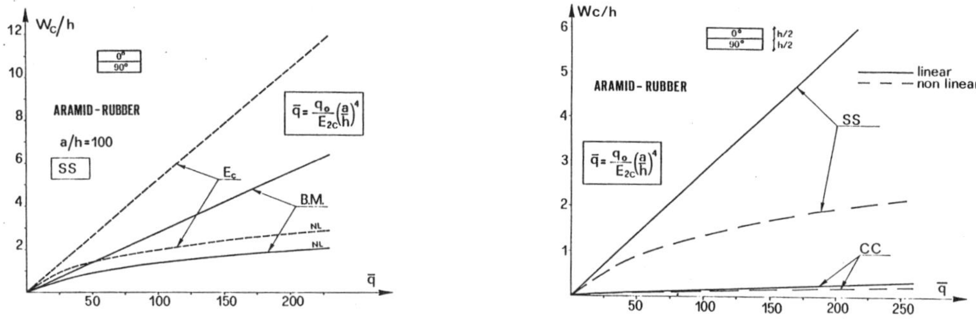

a) b)

Fig. 7 - Cross-ply square plate
a) Influence of material and geometric nonlinearities.
(NL = Geometrical Nonlinearities, B.M. = Bimodulus Material.)
b) Influence of boundary conditions

REFERENCES

[1] *E. Reissner and Y. Stavsky*, Bending and stretching of certain types of heterogeneous aelotropic elastic plates. *J. Appl. Mech. 28, 402-408 (1961).*

[2] *C.W. Bert and B.L. Mayberry*, Free vibrations of unsymmetrically laminated anisotropic plate with clamped edges. *J. Composite materials 3, 282-293 (1969).*

[3] *N.J. Pagano*, Exact solution for composite laminates in cylindrical bending. *J. Composite Materials 3, 398-411 (1969).*

[4] *N.J. Pagano and S.J. Hatfield*, Elastic behavior of multilayer bi-directional composites. *AIAA J., 10, 931-933 (1972).*

[5] *R.M. Jones*, Stress - strain relations for materials with different moduli in tension and compression. *AIAA J., 15, 16-23 (1977).*

[6] *C.W. Bert*, Models for fibrous composites with different properties in tension and in compression. *J. of Eng. Mat. Tech., Trans. of ASME 99H(4), 344- 349, (1977).*

[7] *J.N. Reddy and W.C. Chao*, Non linear bending of thick rectangular, laminated composite plates. *Int. J. Nonlin. Mech. 16, 291-301, (1981).*

[8] *D. Bruno, A. Leonardi and G. Porco,* Non linear analysis of thick sandwich plates. *The second East Asia-Pacific Conf. on Struct. Eng. and Constr. - Chiang Mai, Thailand - 11-13 Jan., 1989.*

[9] *G. Porco, G. Spadea and R. Zinno,* A geometrically nonlinear analysis of laminated composite plates using a shear deformation theory. *Rend. Acc. dei Lincei 1989 (to appear).*

[10] *A. Leonardi and E. Sacco,* Piastre bimodulari su suolo elastico unilaterale. *Atti del IX Congr. Naz. AIMETA, 619-622, Bari, Ott., 1988.*

[11] *C.W. Bert, J.N. Reddy, V. Sudhakar Reddy and W.C. Chao,* Bending of thick rectangular plates laminated of bimodulus composite materials. *AAIA J., 19, 1342-1349 (1981).*

[12] *J.N. Reddy and W.C. Chao,* Finite-element analysis of laminated bimodulus composite - material plates. *Comp. & Struct. 12, 245-251 (1980).*

Domenico Bruno, Giacinto Porco, Raffaele Zinno
Dipartimento di Strutture - Università della Calabria
Arcavacata di Rende (Cosenza) - Italy

International Series of Numerical Mathematics, Vol. 101, © 1991 Birkhäuser Verlag Basel

HOLONOMIC VERSUS RATE RELATIONS IN PLASTICITY

Leone Corradi[1], Francesco Genna[1] and Lorella Annovazzi[2]

[1] Department of Structural Engineering, Politecnico di Milano — Italy
[2] Industrie Magneti Marelli, Milano — Italy

Introduction

The inherently path-dependent nature of plastic behavior requires that the constitutive laws be formulated in terms of rates. The material response to finite stress or strain changes can only be obtained by integrating the rate equations along a given path. Nevertheless, various elastic-plastic laws in finite terms have been proposed, known as *deformation* or *holonomic* theories (among them, those due to Hencky [1] and Nadai [2] are to be mentioned). They must be considered as approximations to the actual *path-dependent* or *non-holonomic* behavior, reasonable only along nearly proportional paths.

In spite of this fact, holonomic elastic-plastic laws are of interest. Some behaviors typical of unilateral problems can be simulated in this way. A notable example is the so-called *no-tension* material, when rupture occurring at zero stress level is conceived as a reversible inelastic deformation. Actually, this behavior exhibits peculiar features which make it worth independent study; its idealization as elastic-plastic and holonomic is perhaps artificial, but legitimate.

In any case, when a computation is performed, rates are replaced by finite increments and within each of them some deformation path is assumed. Along this path the irreversible nature of the material behavior is usually ignored, but the relevant holonomic law rarely is introduced. As a matter of fact, often the path is only implicitly assumed by the solution scheme used and different strategies imply different paths. Their features are investigated a-posteriori, if at all.

It is of obvious importance not only to assess which path the step solution algorithm is equivalent to, but also, whenever possible, to force the solution towards paths enjoying particular properties. Along this line moves the work pioneered by Martin [3-4], aiming at the definition of *extremal* paths, which minimize some suitably defined work function. It can be shown [5] that (i) such paths are ***holonomic***, in the sense that irreversibility plays no role, and (ii) that a sequence of extremal path solutions converges to the rate solution as the step size decreases. In general, solution algorithms based on backward-difference time integration turn out to be equivalent to extremal paths.

Martin's results are important and fairly general. However, explicit expressions for the relevant holonomic constitutive relationship are easily obtained only in some cases, such as Mises condition. This is considered a minor drawback, since the extremal path condition is enforced anyway if a backward-difference algorithm is applied to the rate law. However, confinement to a single solution scheme may be too restrictive, since numerical or convergence problems might suggest that alternative strategies be used. On the other hand, if the equivalent holonomic law is available, not only the step solution becomes independent of the algorithm used, but the computation of the material response often is greatly facilitated. This aspect was demonstrated in [6] by two of the authors, under rather unrestrictive conditions on the yield function.

Scope of this paper is to show that the holonomic laws proposed in [6] are in fact equivalent to extremal paths for rate laws with linear kinematic and isotropic "strain-hardening" (the latter term refers to a generalization of the strain-hardening model, sometimes adopted instead of the work-hardening one for Mises condition [7]). For brevity, attention here is limited to regular yield functions defining an admissible domain closed in stress space. Extensions to a more general context are perhaps not completely straightforward, but can be envisaged. Also, the fully holonomic behavior only is considered; however, generalizations to cover the computationally relevant case in which holonomy is confined within a finite increment entail merely formal complications.

Holonomic Relations

For the case considered here of regular yield functions, the holonomic elastic-plastic relations studied in [6] can be expressed as follows

$$\boldsymbol{\sigma} = \mathbf{D}(\boldsymbol{\varepsilon} - \mathbf{p}) \tag{1}$$

$$\Phi(\boldsymbol{\sigma}, \mathbf{p}, \Lambda) = \psi(\boldsymbol{\sigma} - \mathbf{hp}) - H\Lambda - r; \quad \mathbf{p} = \frac{\partial \Phi}{\partial \boldsymbol{\sigma}}\Lambda \tag{2a, b}$$

$$\Phi \leq 0; \quad \Lambda \geq 0; \quad \Phi\Lambda = 0, \tag{3a - c}$$

where ψ is a *convex function* of "back-stresses" $\boldsymbol{\sigma}-\mathbf{hp}$, $\mathbf{h}$ is a symmetric, positive semidefinite constant matrix, H is a non-negative constant introducing effects analogous to isotropic hardening at the rate level and $r \geq 0$ is the yield stress for the virgin material. Holonomic (reversible) behavior is enforced by Eqs. (3), stating that Λ (and, hence, $\mathbf{p}$) vanishes unless $\Phi = 0$. It was shown that Eqs. (1–3) are always able to provide a unique material response to given strains $\boldsymbol{\varepsilon}$.

The computation of this response is facilitated if additional, but still not unreasonably restrictive, requirements are met. Let the convex function ψ be homogeneous of degree one and suppose that Λ can be expressed as a function of $\mathbf{p}$, differentiable and homogeneous of degree one. Then one can write

$$\psi(\boldsymbol{\sigma} - \mathbf{hp}) = \left[\frac{d\psi}{d(\boldsymbol{\sigma} - \mathbf{hp})}\right]^t (\boldsymbol{\sigma} - \mathbf{hp}) = \left(\frac{\partial \Phi}{\partial \boldsymbol{\sigma}}\right)^t (\boldsymbol{\sigma} - \mathbf{hp}); \tag{4a}$$

$$\Lambda = \omega(\mathbf{p}) = \left(\frac{d\omega}{d\mathbf{p}}\right)^t \mathbf{p}. \tag{4b}$$

The material response is now characterized by the following extremum property

$$\min_{\mathbf{p}} \ \frac{1}{2}(\boldsymbol{\varepsilon} - \mathbf{p})^t \mathbf{D}(\boldsymbol{\varepsilon} - \mathbf{p}) + \frac{1}{2}\mathbf{p}^t \mathbf{hp} + \frac{1}{2}H\omega^2(\mathbf{p}) + r\omega(\mathbf{p}); \tag{5a}$$

$$\text{subject to} \ \ \mathbf{p} \in \mathcal{D}_p; \quad \omega(\mathbf{p}) \geq 0, \tag{5b, c}$$

where $\mathcal{D}_p$ denotes the domain of admissible plastic strains, i.e., the subspace spanned by the normals $\partial\Phi/\partial\boldsymbol{\sigma}$ to the yield surface $\Phi = 0$. Equations (5) are a non-linear constrained optimization problem, poorly suited for direct minimization. Its optimality conditions, however, permit an easy computation of the material response which, in some instances, is produced in closed form [6]. Such a feature makes the proposed law attractive for computational purposes.

A comment on the assumptions Eqs. (4) is in order. Equation (4a) is met by several yield conditions of interest. Equation (4b) is not equally evident; however, in principle, for any $\mathbf{p} \in \mathcal{D}_p$ the value of Λ can be determined from Eq. (2), with the possible exception of particular points such as a cone vertex [6].

Extremal Paths in Rate Plasticity

Consider the following rate plasticity law

$$\dot{\boldsymbol{\sigma}} = \mathbf{D}(\dot{\boldsymbol{\varepsilon}} - \dot{\mathbf{p}}); \quad \varphi(\boldsymbol{\sigma}, \mathbf{p}, \alpha) \leq 0; \tag{6a, b}$$

$$\dot{\mathbf{p}} = \left(\frac{\partial \varphi}{\partial \boldsymbol{\sigma}}\right)\dot{\lambda}; \quad \dot{\lambda} \geq 0; \tag{7a, b}$$

$$\varphi\dot{\lambda} = 0; \quad \dot{\varphi}\dot{\lambda} = 0. \tag{8a, b}$$

The regular yield function φ depends on the current stress, possibly on the plastic strain $\mathbf{p}$ and/or on an isotropic hardening function α. Equations (7) are known as the outward normality rule. Equations (8) account for the irreversible nature of the plastic deformation process, by imposing that the plastic multiplier $\dot{\lambda}$ can be greater than zero only if φ and $\dot{\varphi}$ both vanish. The plastic multiplier $\dot{\lambda}$ plays the same role as Λ in Eq. (3). By way of contrast, however, the quantity $\lambda = \int_0^t \dot{\lambda} d\tau$ is never decreasing.

The isotropic hardening function α is postulated to depend on the plastic strain history, denoted by the vector

$$\boldsymbol{\pi} = \boldsymbol{\pi}(\tau); \quad 0 \leq \tau \leq t,$$

with

$$\mathbf{p} = \boldsymbol{\pi}(t); \quad \dot{\boldsymbol{\pi}} = \dot{\mathbf{p}} \quad \forall \tau.$$

At any istant, the rate of α is proportional to $\dot{\lambda}$

$$\dot{\alpha} = \left(\frac{d\alpha}{d\boldsymbol{\pi}}\right)^t \dot{\mathbf{p}} = \left(\frac{d\alpha}{d\boldsymbol{\pi}}\right)^t \left(\frac{\partial \varphi}{\partial \boldsymbol{\sigma}}\right)\dot{\lambda} = \eta\dot{\lambda}, \tag{9}$$

where the symbol $d\alpha/d\boldsymbol{\pi}$ represents a derivative taken along the path $\boldsymbol{\pi}$, i.e., in the direction of $\dot{\mathbf{p}}$.

If $\varphi = 0$, Eq. (8a) is obviously satisfied. In this case, since φ cannot assume positive values, it must be $\dot{\varphi} \leq 0$, with

$$\dot{\varphi} = \left(\frac{\partial \varphi}{\partial \boldsymbol{\sigma}}\right)^t \dot{\boldsymbol{\sigma}} + \left(\frac{\partial \varphi}{\partial \mathbf{p}}\right)^t \dot{\mathbf{p}} + \frac{\partial \varphi}{\partial \alpha}\dot{\alpha} = \left(\frac{\partial \varphi}{\partial \boldsymbol{\sigma}}\right)^t \dot{\boldsymbol{\sigma}} - \kappa\dot{\lambda}, \tag{10a}$$

where

$$\kappa = -\left[\left(\frac{\partial \varphi}{\partial \mathbf{p}}\right)^t \left(\frac{\partial \varphi}{\partial \boldsymbol{\sigma}}\right) + \eta\frac{\partial \varphi}{\partial \alpha}\right] \tag{10b}$$

is the hardening coefficient, non-negative if Drucker's postulate [5] is complied with.

If $\dot{\varphi}$ vanishes identically along a path, the irreversible nature of the material behavior plays no role. Such a path is *holonomic*.

Consider now the following function of plastic strain history (plastic work density)

$$W(\boldsymbol{\pi}) = \int_0^t \boldsymbol{\sigma}^t(\boldsymbol{\pi})\dot{\mathbf{p}} d\tau. \tag{11}$$

Among all possible paths $\boldsymbol{\pi}$ leading to the same final value $\mathbf{p}$, the one which makes W minimum is called *extremal.* The conditions to be satisfied along it by the rate variables were obtained by Martin and are reported in [5]. In this paper the relations useful in the sequel are summarized, with minor changes dictated by present symbology. The extremal path condition reads

$$\delta W = 0 \quad \forall\, \delta\boldsymbol{\pi} \quad \text{such that } \delta\boldsymbol{\pi}(0) = \delta\boldsymbol{\pi}(t) = \mathbf{0}. \tag{12}$$

From Eq. (11), by performing an integration by parts and accounting for Eq. (7a), one obtains

$$\delta W = \int_0^t (\dot{\mathbf{p}}^t \delta\boldsymbol{\sigma} + \boldsymbol{\sigma}^t \delta\dot{\mathbf{p}}) d\tau = \int_0^t (\dot{\mathbf{p}}^t \delta\boldsymbol{\sigma} - \dot{\boldsymbol{\sigma}}^t \delta\boldsymbol{\pi}) d\tau = \int_0^t \left[\dot{\lambda}\left(\frac{\partial\varphi}{\partial\boldsymbol{\sigma}}\right)^t \delta\boldsymbol{\sigma} - \dot{\boldsymbol{\sigma}}^t \delta\boldsymbol{\pi}\right] d\tau. \tag{13}$$

Along the extremal path it can be assumed, without loss of generality, $\delta\varphi = 0$. In fact, only in this case one can have $\delta\boldsymbol{\pi} \neq \mathbf{0}$. Hence

$$\left(\frac{\partial\varphi}{\partial\boldsymbol{\sigma}}\right)^t \delta\boldsymbol{\sigma} = -\left(\frac{\partial\varphi}{\partial\boldsymbol{\pi}}\right)^t \delta\boldsymbol{\pi} - \frac{\partial\varphi}{\partial\alpha}\delta\alpha = -\left(\frac{\partial\varphi}{\partial\boldsymbol{\pi}} + \frac{\partial\varphi}{\partial\alpha}\frac{d\alpha}{d\boldsymbol{\pi}}\right)^t \delta\boldsymbol{\pi}. \tag{14}$$

Then, the extremal path condition becomes

$$\int_0^t \left[\dot{\boldsymbol{\sigma}} + \left(\frac{\partial\varphi}{\partial\boldsymbol{\pi}} + \frac{\partial\varphi}{\partial\alpha}\frac{d\alpha}{d\boldsymbol{\pi}}\right)\dot{\lambda}\right]^t \delta\boldsymbol{\pi}\, d\tau = 0 \quad \forall\, \delta\boldsymbol{\pi}. \tag{15}$$

When the yield surface defines a bounded domain in stress space, $\delta\boldsymbol{\pi}$ is arbitrary. In this case, the extremal path condition becomes

$$\dot{\boldsymbol{\mu}} = \dot{\boldsymbol{\sigma}} + \left(\frac{\partial\varphi}{\partial\boldsymbol{\pi}} + \frac{\partial\varphi}{\partial\alpha}\frac{d\alpha}{d\boldsymbol{\pi}}\right)\dot{\lambda} = \mathbf{0}. \tag{16}$$

In other instances, the normality rule (7) introduces some constraint on the possible values of $\delta\boldsymbol{\pi}$. For example, in the Mises condition the volumetric component of $\boldsymbol{\pi}$ (and, hence, of $\delta\boldsymbol{\pi}$) must be zero. In this case, Eq. (15) requires that the deviatoric part of $\dot{\boldsymbol{\mu}}$ vanishes.

In any case, the extremal path is holonomic. In fact, by means of straightforward algebra, one can show that

$$\dot{\varphi} = \left(\frac{\partial\varphi}{\partial\boldsymbol{\sigma}}\right)^t \dot{\boldsymbol{\mu}}, \tag{17}$$

which is zero if $\dot{\boldsymbol{\mu}}$ is orthogonal to all admissible plastic strain variations $\delta\boldsymbol{\pi} = (\partial\varphi/\partial\boldsymbol{\sigma})\delta\lambda$, i.e., if Eq. (15) holds.

The Mises condition supplies an example of a regular yield function; in the three-dimensional case and for linear kinematic hardening it reads

$$\varphi = \frac{1}{\sqrt{2}}[(\boldsymbol{\sigma} - c\mathbf{p})^t \mathbf{M}(\boldsymbol{\sigma} - c\mathbf{p})]^{\frac{1}{2}} - F(\alpha) - r, \tag{18}$$

where c is a non-negative constant and the matrix $\mathbf{M}$ is such that $\mathbf{s} = \frac{1}{3}\mathbf{M}\boldsymbol{\sigma}$ is the stress deviator, F represents the isotropic hardening term, where the dependence of the yield function on α is accounted for and, as in Eq. (2a), r is the yield stress for the virgin material. From Eq. (15) then one obtains

$$\dot{\mathbf{s}} - c\dot{\mathbf{p}} - \frac{dF}{d\alpha}\frac{d\alpha}{d\boldsymbol{\pi}}\dot{\lambda} = \mathbf{0}. \tag{19}$$

The left hand side is, as anticipated, the deviatoric part of $\dot{\boldsymbol{\mu}}$. Two alternative choices for the isotropic hardening function α are mostly adopted, namely

Work-hardening:

$$\alpha = W, \tag{20a}$$

where

$$W = \int_0^t \mathbf{s}^t \dot{\mathbf{p}} d\tau \tag{20b}$$

is the plastic work density defined by Eq. (11), in which $\boldsymbol{\sigma}$ is replaced by $\mathbf{s}$ owing to the deviatoric nature of $\dot{\mathbf{p}}$. Then $d\alpha/d\boldsymbol{\pi} = \mathbf{s}$ and Eq. (19) becomes

$$\dot{\mathbf{s}} = c\dot{\mathbf{p}} + \mathbf{s}\frac{dF}{dW}\dot{\lambda}. \tag{21}$$

Strain-hardening:

$$\alpha = \lambda = \int_0^t \dot{\lambda} d\tau, \tag{22a}$$

where

$$\dot{\lambda} = \omega(\dot{\mathbf{p}}) = \frac{\sqrt{2}}{3}(\dot{\mathbf{p}}^t\mathbf{M}\dot{\mathbf{p}})^{\frac{1}{2}} \tag{22b}$$

is the equivalent plastic strain rate. It was denoted by $\dot{\lambda}$ since it does in fact coincide with the plastic multiplier in Eqs. (7) [5]. Then, $d\alpha/d\boldsymbol{\pi} = d\omega/d\dot{\mathbf{p}} = (2/3)(\partial\varphi/\partial\boldsymbol{\sigma})$ (the latter equality flows from Eqs. (18) and (22b)). Thus, Eq. (19) reduces to

$$\dot{\mathbf{s}} = c\dot{\mathbf{p}} + \frac{2}{3}\frac{dF}{d\lambda}\frac{\partial\varphi}{\partial\boldsymbol{\sigma}}\dot{\lambda} = \left(c + \frac{2}{3}\frac{dF}{d\lambda}\right)\dot{\mathbf{p}}. \tag{23}$$

As particular forms of Eq. (19), both the rate relations (21) and (23) hold on the extremal path. Their integration along this path would produce the explicit expression for the relevant holonomic law. Such a result, however, is confined to the three-dimensional Mises condition, an important but not exhaustive example. Extensions to a more general context entail some difficulties in the work-hardening case. On the contrary, isotropic strain-hardening easily undergoes generalization. It will be shown that, when strain-hardening is linear, holonomic laws are produced in the computationally attractive form Eqs. (1–3).

Generalized Linear Strain-hardening Model

Consider the following expression for the yield function φ, Eq. (6b)

$$\varphi = \psi(\boldsymbol{\sigma} - \mathbf{h}\mathbf{p}) - H\lambda - r, \tag{24}$$

where ψ is homogeneous of degree one and, hence, Eq. (4a) holds. As in Eq. (2a), $\mathbf{h}$ is a symmetric, positive semidefinite matrix and H, r are non-negative constants. λ is defined by Eq. (22a) as the integral of the plastic multiplier $\dot{\lambda}$. By analogy with the Mises case, the model governed by Eq. (24) is referred to as generalized strain-hardening, owing to the dependence on λ of the isotropic hardening contribution, linear since the dependence on λ is so.

Assume that $\dot{\lambda}$ is a differentiable and homogeneous of degree one function of $\dot{\mathbf{p}}$, so that one can write

$$\dot{\lambda} = \omega(\dot{\mathbf{p}}) = \left(\frac{d\omega}{d\dot{\mathbf{p}}}\right)^t \dot{\mathbf{p}}. \tag{25}$$

Equation (22b) is an example of such an expression. As recalled in Section 1 with reference to the holonomic law, the relationship (25) can be established in principle for any regular yield function (with the immaterial exception of particular singular points). Note that if $\dot{\mathbf{p}}$ is eliminated from Eq. (25) by making use of normality, Eq. (7a), the following condition, which must hold at solution, is obtained

$$\left(\frac{d\omega}{d\dot{\mathbf{p}}}\right)^t \left(\frac{\partial\varphi}{\partial\boldsymbol{\sigma}}\right) = 1. \tag{26}$$

Attention will be limited to the case in which $\varphi \leq 0$ defines a bounded domain in stress space (plane stress Mises condition is an example). Then, the extremal path condition is given by Eq. (16) which, in the present context, becomes

$$\dot{\boldsymbol{\sigma}} = \mathbf{G}\dot{\mathbf{p}}; \quad \mathbf{G} = \mathbf{h} + H\left(\frac{d\omega}{d\dot{\mathbf{p}}}\right)\left(\frac{d\omega}{d\dot{\mathbf{p}}}\right)^t. \tag{27a, b}$$

In fact, as Eq. (25) shows, $d\omega/d\dot{\mathbf{p}}$ is homogeneous of degree zero and, hence, it is equal to $d\alpha/d\boldsymbol{\pi}$ if α is defined by Eq. (22a) and the derivative is taken along the path. Note that $\mathbf{G}$ is a symmetric and positive semidefinite matrix, since $\mathbf{h}$ is so and $H \geq 0$.

Equations (27) are easily integrated on a ***linear path*** in the plastic strain space. This is defined as one along which the normal vector $\partial\varphi/\partial\boldsymbol{\sigma}$ does not change direction. ψ being homogeneous of degree one, $\partial\varphi/\partial\boldsymbol{\sigma}$ is homogeneous of degree zero and, hence, is constant as long as its direction keeps unchanged. Thus, on a linear path one can write

$$\dot{\mathbf{p}} = \mathbf{n}\dot{\lambda}; \quad \mathbf{n} = \frac{\partial\varphi}{\partial\boldsymbol{\sigma}} = \text{const.} \tag{28a, b}$$

For the same reason, on a linear path $d\omega/d\dot{\mathbf{p}}$ is also constant and the following relation can be written

$$\dot{\lambda} = \mathbf{m}^t\dot{\mathbf{p}}; \quad \mathbf{m} = \frac{d\omega}{d\dot{\mathbf{p}}} = \text{const.} \tag{29a, b}$$

Equation (26) now becomes

$$\mathbf{m}^t\mathbf{n} = 1. \tag{30}$$

As Eq. (29b) shows, matrix $\mathbf{G}$ is constant along a linear path, which makes the integration of Eqs. (27) straightforward. Let $\boldsymbol{\sigma}_0$ denote the stress vector at which yielding is first encountered; one obtains

$$\boldsymbol{\sigma} - \boldsymbol{\sigma}_0 = \mathbf{G}\mathbf{p} = \mathbf{h}\mathbf{p} + H\mathbf{m}\lambda, \tag{31}$$

with

$$\lambda = \mathbf{m}^t\mathbf{p}; \quad \mathbf{p} = \mathbf{n}\lambda. \tag{32a, b}$$

Equation (31) expresses the stress-plastic strain relation along the linear path, which is extremal for the rate law, since Eq. (27a) is complied with at any instant. Note that from this equation it follows that also $\dot{\boldsymbol{\sigma}}$ has constant direction and, hence, because of Eq. (6a), so has $\dot{\boldsymbol{\varepsilon}}$. Moreover, $\mathbf{G}$ being positive semidefinite, one has at any instant $\dot{\boldsymbol{\sigma}}^t\dot{\mathbf{p}} \geq 0$.

Comparison between Eqs. (2a) and (24) brings to light strong similarities between the holonomic and the linear strain-hardening rate models. Actually, for the same material constants and the same expression of ψ, Eq. (4a), the only difference consists of the replacement of Λ, Eq. (4b), with λ, Eq. (22a). The latter is path dependent, while Λ is not so. The function ω is the same for the same yield condition, but integration along arbitrary paths will produce, in general, $\lambda \neq \Lambda$.

It can be immediately recognized, however, that the two values are identical when $\dot{\lambda}$ is integrated along a linear path. In fact, in this case it is also $d\omega/d\mathbf{p} = \mathbf{m}$ and Eq. (4b), account taken of Eqs. (32a) and (30), yields

$$\Lambda = \mathbf{m}^t\mathbf{p} = \mathbf{m}^t\mathbf{n}\lambda = \lambda. \tag{33}$$

The equivalence is also recognized if Eq. (31) is premultiplied by $\mathbf{n}^t$. By rearranging terms and exploiting Eqs. (30) and (31), one obtains

$$\mathbf{n}^t(\boldsymbol{\sigma} - \mathbf{h}\mathbf{p}) - H\Lambda - \mathbf{n}^t\boldsymbol{\sigma}_0 = 0. \tag{34}$$

The first addend is the value of ψ along the linear path, as Eqs. (28b) and (4a) show. Moreover, by definition, $\mathbf{n}^t\boldsymbol{\sigma}_0 = r$. Therefore, Eq. (31) implies that the holonomic yield function Φ is zero, in accordance with positive Λ, Eq. (3), and that the rate solution along the linear path is also a solution of the holonomic relations Eqs. (1–3) which, as mentioned in Section 1, was proved in [6] to be unique.

This proves that the holonomic response coincides with that provided, in the linear strain-hardening case, by the rate law (6–8) integrated along a linear path, which also is an extremal one, i.e., involves the minimum plastic work density among all possible paths leading to the same final plastic strain value. It can be concluded that the holonomic law proposed in [6] (specialized to finite holonomic steps only, if required) is well suited for elastic-plastic computations, in that it couples the operative advantages of an easier evaluation of the material response with a fairly precise assessment of the meaning of the solution obtained.

Acknowledgments

Research supported by the Italian Ministry of Public Education (M.P.I.).

References

[1] HENCKY, H., Zur Theorie Plasticher Deformationen und der Hierdurch in Material Herforgerufener Nachspannungen, *Zeit. Ang. Math. Mech.*, **4**, 323, 1924

[2] NADAI, A. L., Plasticity, McGraw-Hill, New York, 1931

[3] MARTIN, J. B., A Complementary Energy Bounding Theorem for Time-Independent Materials, in Frederick, D. and Harris, E. H. (editors), Developments in Theoretical and Applied Mechanics, Pergamon Press, New York, 517, 1970

[4] PONTER, A. R. S. and MARTIN, J. B., Some Extremal Properties and Energy Theorems for Inelastic Materials and their Relationships to the Deformation Theories of Plasticity, *J. Mech. Phys. Solids*, **20**, 281, 1972

[5] MARTIN, J. B., Plasticity: Fundamentals and General Results, MIT Press, Cambridge, Mass., 1975

[6] CORRADI, L. and GENNA, F., Kinematic Extremum Theorems for Holonomic Plasticity, *Int. J. Plasticity*, to appear

[7] MENDELSON, A., Plasticity: Theory and Application, Mac Millan Co., New York, 1968

Leone Corradi, Francesco Genna
Department of Structural Engineering
Politecnico di Milano
Piazza Leonardo Da Vinci, 32
20133 Milano — Italy

Lorella Annovazzi
Metodologie di Progettazione
Industrie Magneti Marelli
Via Adriano, 81
20128 Milano Crescenzago — Italy

International Series of Numerical Mathematics, Vol. 101, © 1991 Birkhäuser Verlag Basel

A UNIFIED TREATMENT OF DAMAGE AND PLASTICITY BASED ON A NEW DEFINITON OF MICROFRACTURE

Gianpietro Del Piero[1] and Rubens Sampaio[2]

[1] Istituto di Meccanica Teorica ed Applicata, Università di Udine, Udine, Italy;
[2] Departamento de Engenharia Mecânica, Pontificia Universidade Catòlica do Rio de Janeiro, Rio de Janeiro, Brasil.

Abstract. Using a new concept of configuration, we show how to treat in a unified way, in one dimension and under tensile forces, plastic deformation, damage, or a combination of both.

1. Introduction

One of the basic aims of Mechanics is the prediction of whether or not a continuous body or a structure can sustain a prescribed set of external loads without fracturing. In this respect, Fracture Mechanics describes with good accuracy the propagation inside a body of an existing crack, but fails in answering the fundamental question of how and when a crack does appear. A central difficulty in modeling fracture phenomena is that they involve, by their own nature, discontinuous displacements; therefore, they violate one of the basic assumptions of Continuum Mechanics, the continuity and differentiability of the displacement fields. A related difficulty is that the presence of fractures induces unilateral restrictions both on the stress and on the discontinuities of the displacement.

An alternative approach to the problem of rupture is supplied by continuum theories. It originates from the observation that "failure of many materials involves propagation of a large system of densely distributed cracks rather than a single, precisely defined fracture.... The number of cracks or microcracks is extremely large, and their locations and orientations are random. Therefore, it is unavoidable to treat the densely cracked material as a continuum" [1]. Continuum theories have been successful in describing such phenomena as plasticity and damage [2],[3]. Two remarks, however, must be made. The first one is that the existing theories do not cover the case in which distributed microcracks coexist with macroscopic fractures; the second is that continuum theories have so far been successful in providing satisfactory models only in the one-dimensional case.

When studying materials which do not support tension, using in part ideas developed in discussions with D. R. Owen, one of us was led to develop a sort of extended kinematics of continua, allowing for discontinuous displacements [4],[5]. It came out that a proper choice of the function space defining the set of admissible displacements led to describe microfractures as limit elements of sequences of discontinuous displacement fields, in which the discontinuities converge to zero in some suitable norm. As it will be explained in the next Section, one naturally is led to a generalized notion of configuration, in which a configuration is a triple consisting of a displacement field and of two more fields which describe the state of microscopic fracture of the material. In the existing continuum theories, this supplementary information is obtained by specifying a number of "state variables" [2],[6]; in the present approach, it comes directly from kinematics.

In this paper we deal only with one-dimensional bodies subject to a tensile axial force. In this restricted context, we use the generalized concept of configuration to construct a model of material response which takes into account both plastic deformation and damage. These phenomena are schematized as microfractures obeying two different constitutive laws. The resulting force-elongation behaviour coincides with the one predicted by some well known phenomenological theories. There is some hope of extending this model to more dimensions, contributing thereby to the solution of one of the main open problems in the theory of inelastic material behaviour: the formulation of the incremental equilibrium problem for a three-dimensional body.

2. Configurations

In one dimension, macrofractured configurations are described by displacement functions which are continuous and continuously differentiable, except at a finite number of points. More precisely, we say that a real-valued function v defined over the interval (a,b) is piecewise H^1 if (a,b) can be subdivided into a finite number n+1 of subintervals (x_i, x_{i+1}), with $i=0,1,2,..n$, $n=n(v)$, $x_0=a$ and $x_{n+1}=b$, such that v is H^1 in each subinterval.

It follows from well known theorems in Sobolev spaces that v is continuous in each open subinterval (x_i,x_{i+1}) and has a continuous extension to its closure. In particular, denoting by $[v(x_i)]$, the discontinuity of v at x_i, i= 1,2,..n, and by $[v(x_0)]$ and $[v(x_{n+1})]$ the traces of v at x_0 and x_{n+1}, we have that the sum

$$\sum_{i=0}^{n+1} [v_i]^2$$

is finite. Note also that, because the set of the discontinuity points has Lebesgue measure zero, v is a function in $L^2(a,b)$ and its distributional derivative Dv is a measure whose absolutely continuous part $D^a v$ is L^2 over (a,b).When interpreting v as the displacement function of a one-dimensional body from a non-fractured reference configuration, the x_i, i=1,2,..n, are the points at which a fracture occurs, and the subintervals (x_i,x_{i+1}) are the parts of the body which remain intact. Clearly, at the interior of each of these subintervals the regularity of the displacement is the same as in the ordinary continuum.

The set of all macrofractured configurations is a vector space. On it, we introduce the norm

$$|||\, v\, ||| := \Big(\| v \|^2_{L^2(a,b)} + \| D^a v \|^2_{L^2(a,b)} + \sum_{i=0}^{n(v)+1} [v(x_i)]^2 \Big)^{1/2}. \tag{2.1}$$

It can be verified that this space is not complete. Indeed, taking a Cauchy sequence $\{v_k\}$, we have that $\{v_k\}$ converges to some f and $\{D^a v_k\}$ converges to some F in $L^2(a,b)$, but F is not equal, in general, to $D^a f$. For example, it is shown in [5] that the sequence of the piecewise constant functions

$$v(x) := \frac{h}{k}, \qquad \frac{h-1}{k}(b-a) < x < \frac{h}{k}(b-a), \qquad h = 1,2,...k, \tag{2.2}$$

is Cauchy with respect to the norm (2.1) and converges in L^2 to the function $f(x)=x$, whereas $D^a v=0$ implies $F(x)=0$ and therefore $D^a f \neq F$.

With a similar example [5] it can be shown that the collection of the discontinuities of v_k

$$Jv_k := \{ [v_k(x_i)],\ i=1,2,..n(v_k) \} \tag{2.3}$$

converges to a finite or countable collection $\Phi := \{ [v(x_i)],\ i=1,2,..n(v) \}$ of discontinuities which is, in general, different from the collection of the discontinuities of f. In conclusion, the limit element of a Cauchy sequence of macrofractured configurations can be identified with a triple

$$v := \{ f, F, \Phi \} , \tag{2.4}$$

with f and F in $L^2(0,l)$ and Φ a finite or countable collection of localized discontinuities in l^2. A triple of the type (2.4) will be called a configuration. Macrofractured configurations are triples for which $F=D^a f$ and for which f admits a collection of discontinuities Jf such that $\Phi = Jf$. For such configurations, $D^a f$ measures the elastic deformation and Jf is the collection of the macrofractures. A microfractured configuration is a triple for which at least one of the above equalities fails to hold. The fact that each microfractured configuration v is the limit element of a sequence $\{v_k\}$ of macrofractured configurations with $D^a v_k \to F$ and $Jv_k \to \Phi$ suggests that F has to be identified with the elastic deformation and Φ with the collection of the macrofractures of v. The function $D^a f - F$ is interpreted as a density of diffused microfractures and, whenever Jf can be defined, $Jf - \Phi$ is interpreted as a collection of concentrated microfractures.

3. Constitutive assumptions

A macrofracture is considered as a macroscopic separation of the material. Consequently, any discontinuity $[v(x_i)]$ in the collection Φ must be positive, in order to avoid interpenetration of matter. Moreover, we assume that there is no transmission of stress across a macrofracture. Denoting by $T(x_i)$ the stress at x_i, we formulate our first constitutive assumption as follows:

$$[v(x_i)] > 0 , \qquad T(x_i) = 0 . \tag{3.1}$$

Microfractures are assumed to model phenomena at the microscopic scale, such as plastic yielding or damage, which do not prevent the transmission of stress. For microfractures, negative discontinuities are allowed.

The second constitutive assumption is that the non-fractured material is linear elastic:

$$T = c\, F, \tag{3.2}$$

with c a positive constant. The third assumption is that, at almost all points x in (a,b) and for any time in $[0,\Theta]$ the elastic deformation is determined in the following way. Write, for convenience, H instead of $D^a f$, and define a state of the material at x as a pair (H,F) and a deformation process at x as a continuous one-parameter family $H=\hat{H}(t)$, $0\le t\le\Theta$. We assume that for any initial state (H_0,F_0) and for any deformation process $\hat{H}$ with $\hat{H}(0)=H_0$ the elastic deformation $F=\hat{F}(t)$ is determined as the solution of the initial-value problem

$$(3.3)\qquad \frac{d}{dt}\hat{F}(t) = \boldsymbol{f}(\hat{H}(t), \hat{F}(t), \frac{d}{dt}\hat{H}(t)), \qquad \hat{F}(0) = F_0.$$

Because H is the macroscopic displacement gradient and F is proportional to the stress by (3.2), this assumption characterizes the material as a material of the rate type. The structure of the incremental response function $\boldsymbol{f}$ depends upon the particular phenomena to be modeled. In the following Section, we show that some classical theories of damage and plastic flow can be obtained by means of an appropriate choice of $\boldsymbol{f}$.

Our last assumption is that $\boldsymbol{f}$ is positively homogeneous of degree one:

$$(3.4)\qquad \boldsymbol{f}(H,F,\lambda\dot{H}) = \lambda\ \boldsymbol{f}(H,F,\dot{H}) \qquad \forall\lambda\ge 0$$

(here and in the following we denote time derivatives by a superimposed dot). It is known that this assumption guarantees that the material is rate-independent, i.e., that the solution of (3.3) is invariant under monotonic changes of the time scale. In particular, when the initial state $(\hat{H}(0),F_0)$ is the natural state (0,0) and $\hat{H}$ is non-decreasing, the state-process pair is called loading and the curve $F=\tilde{F}(H)$ in the (H,F) plane determined by (3.3) is called the loading curve. Unloading is a state-process pair in which the state is a point $(H, \tilde{F}(H))$ of the loading curve and the process starts from $\tilde{F}(H)$ and has decreasing values. Finally, reloading is a pair in which the state (H,F) does not belong to the loading curve and the process starts from F and is increasing.

4. Incremental laws

In this Section we show how a proper choice of the incremental response function allows reproducing some known models in damage and plasticity. A combined damage-plasticity model is also obtained. For sim= plicity, we consider only states characterized by positive values of F and H.

(i) Damage. In the present approach in terms of generalized configurations, we assume as measure of damage the ratio between the microfracture density H-F and the macroscopic displacement gradient H:

$$\rho := \frac{H-F}{H}. \tag{4.1}$$

Our constitutive assumption is that ρ grows with constant rate during loading, and remains constant during unloading and reloading:

$$\dot{\rho} = \begin{cases} \alpha \dot{H} & \text{at loading,} \\ 0 & \text{at unloading and reloading,} \end{cases} \tag{4.2}$$

where α is a positive material constant. In terms of (4.1), the loading law becomes

$$\dot{F} = \left(\frac{F}{H} - \alpha H\right) \dot{H}, \tag{4.3}$$

or, in view of the assumption of rate-independence (3.4),

$$\frac{dF}{dH} = \frac{F}{H} - \alpha H, \tag{4.4}$$

and integration yields

$$F = kH - \alpha H^2. \tag{4.5}$$

The initial condition F=0 at H=0 is automatically satisfied. To determine the integration constant k we evaluate ρ from (4.1) and(4.5)

$$\rho = 1 - k + \alpha H, \tag{4.6}$$

and we require that ρ=0 at H=0, i.e., that the material at the natural state has no damage. It comes out that k=1 and, therefore,

$$F = \tilde{F}(H) = H - \alpha H^2. \tag{4.7}$$

This is the equation of the loading curve. By (4.2), the incremental law for unloading and reloading is deduced simply by putting α=0 in (4.3). It is then immediate to see that the explicit form of $\boldsymbol{f}$ is

$$\boldsymbol{f}(H,F,\dot{H}) = \begin{cases} \left(\frac{F}{H} - \alpha H\right) \dot{H} & \text{for } F= H-\alpha H^2 \text{ and } \dot{H}>0, \\ \frac{F}{H} \dot{H} & \text{otherwise.} \end{cases} \tag{4.8}$$

This constitutive law corresponds to the "standard material of Continuum Damage Mechanics", see Fig. 1a. In [7], this model is obtained by considering damage as due to the presence of microvoids, and by assuming a linear relationship between the macroscopic strain and the "conventional stress", i.e., the stress evaluated on the area not occupied by the voids. Here, the same law has been deduced from the assumption (4.2) which governs the growth of the density of the diffused microfractures.

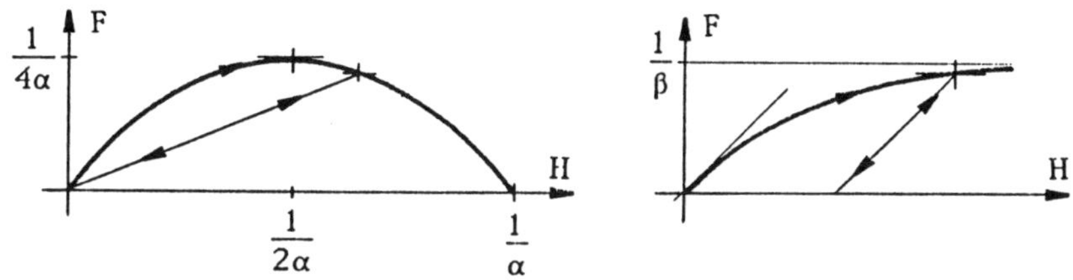

Fig.1

(a) Damage model (b) Plasticity model

(ii) Plasticity. In the elastic-plastic case, the microfracture density H-F is considered as responsible for the plastic deformation. Here we consider a model without yield condition, i.e., a model in which plastic deformation begins as soon as the material leaves the natural state. More precisely, we assume that

$$(4.9) \qquad (H-F)^{\cdot} = \begin{cases} \beta\, F\, \dot{H} & \text{for monotonic loading,} \\ 0 & \text{otherwise,} \end{cases}$$

where β is a positive material constant. At loading we get

$$(4.10) \qquad \dot{F} = (1 - \beta F)\, \dot{H} ,$$

or

$$(4.11) \qquad \frac{dF}{dH} = 1 - \beta F ,$$

and, after imposing the initial condition F=0 at H=0,

$$(4.12) \qquad F = \beta^{-1} (1-e^{-\beta H}).$$

At unloading, from (4.10) with β=0 we get $\dot{F} = \dot{H}$. Hence, the incremental response function is

$$(4.13) \qquad \boldsymbol{f}(H,F,\dot{H}) = \begin{cases} (1-\beta F)\, \dot{H} & \text{for } F=\beta^{-1}(1-e^{-\beta H}) \text{ and } \dot{H}>0, \\ \dot{H} & \text{otherwise.} \end{cases}$$

It corresponds to the model depicted in Fig.1b.

(iii) Combined damage and plasticity. In this model the total deformation H is assumed to be the sum of an elastic part F, a plastic part P and a damage part D:

(4.14) $$H = F + P + D.$$

The measure of damage is defined as

(4.15) $$\rho = \frac{D}{H},$$

and P is the measure of plastic flow. ρ and P are assumed to obey the incremental laws (4.2) and (4.9), respectively:

(4.16) $$\begin{cases} \left(\frac{D}{H}\right)^{\cdot} = \alpha\dot{H} & \text{and} \quad \dot{P} = \beta\, F\, \dot{H} \quad \text{at loading,} \\ \left(\frac{D}{H}\right)^{\cdot} = 0 & \text{and} \quad \dot{P} = 0 \quad \text{at unloading and reloading.} \end{cases}$$

In order to determine the response function ***f***, let us first consider loading. Proceding as in the case (i) and requiring again ρ=0 at the natural state, in analogy with (4.7) we find

(4.17) $$D = \alpha H^2.$$

Moreover, recalling that, from (4.16) and (4.14),

(4.18) $$\frac{dP}{dH} = \beta F = \beta\,(H - D - P),$$

and using (4.17), by integration we get

(4.19) $$P = k\, e^{-\beta H} + \frac{\beta+2\alpha}{\beta^2}(\beta H - 1) - \alpha H^2.$$

The integration constant k is determined by imposing P=0 at H=0:

(4.20) $$k = \frac{\beta+2\alpha}{\beta^2}.$$

Recalling that $F = \beta^{-1}\, dP/dH$ by (4.16), from differentiation of (4.19) we get

(4.21) $$F = \tilde{F}(H) = \frac{\beta+2\alpha}{\beta^2}(1 - e^{-\beta H}) - \frac{2\alpha}{\beta} H.$$

This is the equation of the loading curve. By inspection, we see that there is an interval $(0, H_p)$ in which $\tilde{F}(H)$ is positive. In the same interval, D and dP/dH are positive by (4.17) and (4.18), respectively. Because P=0 at H=0, we deduce that, at loading, the plastic deformation P is positive and grows with H, as far as $H < H_p$.

Let us examine unloading starting from some point (H_o, F_o) of the loading curve, with $F_o = \tilde{F}(H_o)$. The corresponding values D_o, P_o of D and P are given by (4.17) and (4.14), respectively. By the unloading laws (4.16) it follows at once that

$$\frac{D}{H} = \frac{D_o}{H_o} = \alpha H_o , \qquad P = P_o = H_o(1 - \alpha H_o) - F_o . \tag{4.22}$$

Using again (4.14),

$$F = H - P - D = H - H_o(1 - \alpha H_o) + F_o - \alpha H_o H = F_o + (H - H_o)(1 - \alpha H_o). \tag{4.23}$$

This is the equation of the unloading curve. We see that in the present

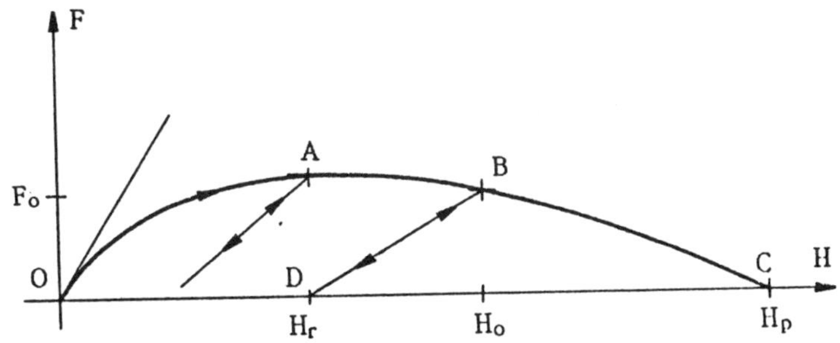

Fig.2.
Coupled damage-plasticity model.
OABC = loading curve; BD = unloading line from B;
OD = residual deformation after unloading from B.

model the unloading curves are straight lines crossing the H-axis at the points

$$H_r := H_o - \frac{F_o}{1 - \alpha H_o} = \frac{P}{1 - \alpha H_o} . \tag{4.24}$$

H_r is the <u>residual strain</u> at complete unloading. An elementary computation shows that H_r grows monotonically with H_o for H_o in $(0, H_p)$. Indeed, the differentiation of (4.24) yields

$$\frac{dH_r}{dH_o} = \frac{1}{1 - \alpha H_o} \frac{dP}{dH_o} + \frac{\alpha}{(1 - \alpha H_o)^2} P ; \tag{4.25}$$

the positiveness of P and dP/dH_0 in $(0,H_p)$ have already been established, and that of $(1-\alpha H_0)$ is an immediate consequence of $(4.22)_4$. The incremental response function can now be easily determined from the identities

$$(4.26) \qquad \boldsymbol{f}(H,F,\dot{H}) = \dot{F} = (1 - \frac{dD}{dH} - \frac{dP}{dH})\,\dot{H}\,,$$

by using (4.17),(4.18) for loading and (4.22) for unloading. The final result is:

$$(4.27) \qquad \boldsymbol{f}(H,F,\dot{H}) = \begin{cases} (1-\ 2\alpha H-\ \beta F)\ \dot{H} & \text{for } F=\tilde{F}(H) \text{ and } \dot{H}>0, \\ (1-\ 2\alpha H_0)\ \dot{H} & \text{otherwise,} \end{cases}$$

where $F=\tilde{F}(H)$ is the equation (4.21) of the loading curve. The model exhibits the behaviour shown in Fig.2. Notice that the expression of $\boldsymbol{f}$ at unloading is not explicit. Indeed, given any point in the (H,F) plane lying below the loading curve, the determination of the corresponding H_0 requires the solution of the system (4.21),(4.23) in the unknowns F_0, H_0.

5. The incremental equilibrium problem for homogeneous configurations

Consider a bar (a,b) subject to a tensile force $\sigma>0$ at the ends. If the bar is of constant cross section then, by equilibrium,

$$(5.1) \qquad T(x) = \sigma\,, \qquad a < x < b,$$

and therefore, by (3.2), the elastic deformation F is the same at any x. Moreover, from T>0 and (3.1) it follows that there are no discontinuities $[v(x_i)]$, i.e., that $\Phi=0$. If it happens that H is also independent of x, we say that the equilibrium configuration is <u>homogeneous</u>. Thus, a homogeneous configuration is characterized by a pair (H,F) of real numbers, with F non-negative. We also say that the bar is <u>homogeneous</u> if the incremental response function does not depend explicitly upon x.

Assume that a homogeneous bar is in equilibrium in a homogeneous configuration (H,F) under the external force σ. The <u>incremental equilibrium problem</u> consists in determining the new equilibrium configuration corresponding either to a perturbation $\dot{\sigma}$ of the external force or to a perturbation $\dot{\delta}$ of the relative displacement between the ends of the bar. The following Proposition states that the invertibility of the function $\boldsymbol{f}(H,F,\cdot)$ implies that the new equilibrium configuration is homogeneous, independently of the particular form of $\boldsymbol{f}$. This result is important because it ensures that, when the initial equilibrium configuration is homogeneous and $\boldsymbol{f}(H,F,\cdot)$ is invertible, the incremental response function not only governs the pointwise stress-strain relationship, but also the global force-elongation behaviour of the bar.

Proposition. Let (H,F) be a homogeneous configuration of a homogeneous bar, in equilibrium under the tensile force σ. Let, further, $\boldsymbol{f}(H,F,\cdot)$ be invertible. Then for any prescribed perturbation either of the force or of the relative displacement of the ends, there is a unique solution to the incremental equilibrium problem. Moreover, the new equilibrium configuration is homogeneous.

Proof. (i) Let $\dot{\sigma}$ be a perturbation of the external force. By (5.1), the corresponding changes in the stress and in the elastic deformation are

(5.2) $$\dot{T} = \dot{\sigma} \quad , \quad \dot{F} = c\,\dot{\sigma} \ ,$$

respectively. If $\boldsymbol{f}(H,F,\cdot)$ is invertible, there is a function $\boldsymbol{g}$ such that

(5.3) $$\dot{H} = \boldsymbol{g}(H,F,\dot{F}) \ ,$$

and, because H,F and $\dot{F}$ are all independent of x, it follows that $\dot{H}$ is determined uniquely by (5.3) and is independent of x.

(ii) Let $\delta = H(b-a)$ be the relative displacement of the ends of the bar in the homogeneous configuration (H,F), and let $\dot{\delta}$ be a given perturbation of δ. In this case, $\dot{T}$ and $\dot{F}$ are not known, but they are certainly independent of x, because the equations (5.2) still hold and $\dot{\sigma}$ is independent of x. By (5.3), $\dot{H}$ is independent of x as well, and the perturbed equilibrium configuration is homogeneous. Moreover, $\dot{H}$ is determined by the geometric condition $\dot{H}(b-a) = \dot{\delta}$, and $\dot{F}$ and $\dot{T}$ follow from $\dot{F} = \boldsymbol{f}(H,F,\dot{H})$ and from (5.2). □

In the examples worked out in Sect.4 the function $\boldsymbol{f}(H,F,\cdot)$ is invertible, except at the non-increasing branch of the loading curve. For homogeneous configurations represented by a point located on this branch, a variety of behaviours is possible. For example, for prescribed $\dot{\sigma}$ we see from (5.2) that $\dot{\sigma}>0$ implies $\dot{F}>0$, and this is incompatible with the incremental law (4.31). This situation is interpreted as describing the appearance of macro= fractures. On the other hand, for $\dot{\sigma}<0$ the incremental law admits two solutions, say $\dot{H}_1$ and $\dot{H}_2$, one along the loading curve and the other one along the unloading curve. In this case, the perturbed equilibrium configu= ration is not expected to be homogeneous. What is expected is the coexistence of two different "phases", characterized by the elastic defor= mation $\dot{H}_1$ and $\dot{H}_2$, respectively. Accordingly, the change in the relative distance of the ends is given by

$$\dot{\delta} = \lambda \dot{H}_1(b-a) + (1-\lambda) \dot{H}_2(b-a) , \tag{5.4}$$

where λ, the proportion of the two phases, can be any real in $(0,1)$. Therefore, $\dot{\delta}$ can take any value between $\dot{H}_1(b-a)$ and $\dot{H}_2(b-a)$.

Acknowledgements

This research has been done within a bilateral project supported by the National Research Councils of Italy (CNR) and Brasil (CNPq). Also acknowledged are the partial supports of the Italian Ministry for Education (MPI) for the first author, and of the Alexander von Humboldt Stiftung for the second author.

References

[1] BAZANT, Z.P., *Mechanics of Distributed Cracking*, Appl. Mech. Reviews, Vol. 39, p. 675-705, 1986.

[2] KRAJCINOVIC, D., *Continuum Damage Mechanics*, Appl. Mech. Reviews, Vol. 37, p. 1-6, 1984.

[3] DRUCKER, D.C., *Conventional and Unconventional Plastic Response and Representation*, Appl. Mech. Reviews, Vol. 41, p. 151-167, 1988.

[4] DEL PIERO, G. *Recent Developments in the Mechanics of Materials which do not Support Tension*, Proc. Internat. Colloquium on Free Boundary Problems, Irsee, W. Germany, 1987 (to appear).

[5] DEL PIERO, G., *A New Function Space for the Mathematical Theory of Plasticity*, Proc. Symp. *Plasticity '89*, Tsu, Japan, 1989, A.S. KHAN & M. TOKUDA Editors., Pergamon Press 1989.

[6] LEMAITRE, J., & J. L. CHABOCHE, *Mécanique des matériaux solides*, Dunod, 1985.

[7] JANSON, J., & J. HULT, *Fracture Mechanics and Damage Mechanics, a Combined Approach*, Journ. Méc. Appl., Vol.1, p. 69-84, 1977.

Gianpietro Del Piero
Università di Udine
Istituto di Meccanica Teorica ed Applicata
viale Ungheria 43, 33100 Udine, Italy

Rubens Sampaio
Pontificia Universidade Catòlica do Rio de Janeiro
Departamento de Engenharia Mecânica
Rua Marques de São Vicente 225, 22543 Rio de Janeiro, Brasil

SHAPE OPTIMIZATION IN UNILATERAL BOUNDARY VALUE PROBLEMS

J. HASLINGER

KFK MFF UK, Malostranské 2/25
11800 Praha 1, Czechoslovakia

1. Introduction

The aim of this contribution is to present some results, concerning the shape optimization in unilateral boundary value problems. In classical (direct) problems, one looks for the response of a structure subjected to given forces (for details see e.g. [5]). In optimal shape design problems, we try to identify a shape of a structure, response of which is as close as possible to designer's ideas. In contact problems one of the possible required criteria is to find a shape of contact surfaces along which stresses are distributed without undesirable stress concentration zones. Question arises, which cost functional guarantee such phenomena. In [4] authors used the total potential energy evaluated in the equilibrium state. Restricting ourselves to simple Dirichlet–Signorini problem for a scalar function we show that under some conditions the minimization of the total potential energy really leads to a constant distribution of a flux along contact part. Moreover the whole problem is smooth, regardless the mapping: control $\mapsto$ state is only directionally differentiable. Shape optimization for contact problems, including approximation results and numerical realization is widely discussed in [1].

1. Setting of the problem

Let $\Omega \subset \mathbf{R}^2$ be a bounded domain with Lipschitz boundary $\partial\Omega = \bar\Gamma_1 \cup \bar\Gamma_2$, where

$$\Omega \equiv \Omega(\alpha) = \{(x_1,x_2) \in \mathbf{R}^2 \mid 0 < x_1 < \alpha(x_2)\ ,\ x_2 \in (0,1)\},$$
$$\Gamma_2 \equiv \Gamma_2(\alpha) = \{(x_1,x_2) \in \mathbf{R}^2 \mid x_1 = \alpha(x_2)\ ,\ x_2 \in (0,1)\}.$$

$\alpha : [0,1] \mapsto \mathbf{R}^1$ is a Lipschitz continuous function, describing a variable part of $\partial\Omega$ and playing the role of a design variable. On $\Omega(\alpha)$ we assume the following unilateral boundary value problem:

$$(\mathcal{P}'(\alpha))_0 \qquad \begin{cases} -\Delta u = f & \text{in } \Omega(\alpha), \\ u = 0 & \text{on } \Gamma_1, \\ u \geq 0,\ \partial u/\partial n \geq 0,\ u\partial u/\partial n = 0 \text{ on } \Gamma_2(\alpha), \end{cases}$$

i.e. on a variable part of $\partial\Omega$ we assume classical Signorini type conditions. Variational form of $(\mathcal{P}'(\alpha))_0$ is given by

$$(\mathcal{P}(\alpha))_0 \qquad \begin{cases} \text{Find } u_0 \equiv u_0(\alpha) \in K(\Omega(\alpha)) \text{ such that} \\ (\nabla u_0, \nabla(v-u_0))_{0,\Omega(\alpha)} \geq (f, v-u_0)_{0,\Omega(\alpha)} \quad \forall v \in K(\Omega(\alpha)), \end{cases}$$

where

$$K(\Omega(\alpha)) = \{v \in H^1(\Omega(\alpha)) \mid v = 0 \text{ on } \Gamma_1,\ v \geq 0 \text{ on } \Gamma_2\}.$$

$(\ ,\ )_{0,\Omega(\alpha)}$ stands for the scalar product in $L^2(\Omega(\alpha))$ and $f \in L^2(\hat{\Omega})$, where $\hat{\Omega}$ is some larger domain, containing $\Omega(\alpha)$ for all α. It is known that $(\mathcal{P}(\alpha))_0$ has exactly one solution u_0, which can be equivalently characterized as a minimizer of

$$J_0(v) = \frac{1}{2}\|\nabla v\|^2_{0,\Omega(\alpha)} - (f, v)_{0,\Omega(\alpha)} \tag{1.1}$$

over $K(\Omega(\alpha))$ and $\|\ \ \|_{0,\Omega(\alpha)} = (\ ,\ )^{\frac{1}{2}}_{0,\Omega(\alpha)}$.

Now we specify the class of admissible variations of α. To this end we define:

$$\begin{aligned} U_{\text{ad}} = \{\alpha \in C^{0,1}([0,1]) \mid 0 < C_0 \leq \alpha(x_2) \leq C_1 \qquad & \forall x_2 \in [0,1], \\ |\alpha(x_2) - \alpha(\bar{x}_2)| \leq C_2|x_2 - \bar{x}_2| \qquad & \forall x_2, \bar{x}_2 \in [0,1], \\ \operatorname{meas}\Omega(\alpha) = C_3\}. & \end{aligned}$$

C_0, C_1, C_2, C_3 are positive constants chosen in such a way that $U_{\text{ad}} \neq \emptyset$. U_{ad} contains functions, which are uniformly bounded, uniformly Lipschitz continuous and preserve the area of $\Omega(\alpha)$. Optimal shape design problem will be stated as follows:

$$(\mathsf{P}) \qquad \begin{cases} \text{Find } \alpha^* \in U_{\text{ad}} \text{ such that} \\ E(\alpha^*) \leq E(\alpha) \qquad \forall \alpha \in U_{\text{ad}}, \end{cases}$$

where $E(\alpha) = J_0(u_0(\alpha))$, i.e. $E(\alpha)$ is equal to the value of the total potential energy evaluated in the equilibrium state $u_0(\alpha)$. Using results of [**1**], one has

THEOREM 1.1. *If $U_{\text{ad}} \neq \emptyset$, then* (P) *has at least one solution α^*.*

2. Interpretation of (P)

In this part we show that our choice of the cost functional really leads to the required effect, namely to the constant distribution of $\partial u/\partial n$ along the part of $\Gamma_2(\alpha^*)$, where the contact is realized. To this end we shall calculate the derivative of E with respect to design variable α.

Let $V = (V_1, 0)$ be a vector field, $V_1 \in H^1(\Omega(\alpha)), V_1(0, x_2) = 0$ and assume a mapping $F_t : \mathbb{R}^2 \mapsto \mathbb{R}^2$ given by

$$F_t(x_1, x_2) = (x_1, x_2) + t(V_1(x_1, x_2), 0), \quad t > 0, \quad (x_1, x_2) \in \overline{\Omega}(\alpha).$$

Denote by $\Omega_t(\alpha) = F_t(\Omega(\alpha))$, $\Gamma_{2t}(\alpha) = F_t(\Gamma_2(\alpha))$ and by $E_t(\alpha) = J_t(u_t(\alpha))$, where $u_t(\alpha) \in K(\Omega_t(\alpha))$ is a solution of the state problem $(\mathcal{P}(\alpha))_t$ (assumed on $\Omega_t(\alpha)$) and J_t is the corresponding total potential energy, evaluated over $\Omega_t(\alpha)$. Our aim will be to calculate

$$\dot{E}(\alpha) \equiv \lim_{t \to 0+} \frac{J_t(u_t) - J_0(u_0)}{t}.$$

Before doing that we reformulate $(\mathcal{P}(\alpha))_t$, $t \geq 0$, introducing Lagrangian multipliers to release the unilateral boundary constraint $u_t \geq 0$ on $\Gamma_{2t}(\alpha)$. Let $L^2_+(\Gamma_{2t}(\alpha))$ denote the set of non–negative square integrable functions on $\Gamma_{2t}(\alpha)$. Then

$$v \in K\left(\Omega_t(\alpha)\right) \Longleftrightarrow v \in H^1(\Omega_t(\alpha)),\ v = 0 \text{ on } \Gamma_1 \text{ and } \int_0^1 \mu v\, dx_2 \geq 0$$

for any $\mu \in L^2_+(\Gamma_{2t}(\alpha))$ (here $\Omega_0(\alpha) \equiv \Omega(\alpha)$).

Denote by $\mathcal{U}_t(\alpha) = \{v \in H^1(\Omega_t(\alpha)) \mid v = 0 \text{ on } \Gamma_1\}$.

It is easy to see that

$$\inf_{v \in K(\Omega_t(\alpha))} J_t(v) = \inf_{v \in \mathcal{U}_t(\alpha))} \sup_{\mu \in L^2_+(\Gamma_{2t}(\alpha))} \left\{ J_t(v) - \int_0^1 \mu v\, dx_2^t \right\},$$

where

$$\int_0^1 \mu v\, dx_2^t \stackrel{\text{def}}{\equiv} \int_0^1 \mu\left(\alpha(x_2) + tV_1\left(\alpha(x_2), x_2\right), x_2\right) v\left(\alpha(x_2) + tV_1\left(\alpha(x_2), x_2\right), x_2\right) dx_2.$$

Denote by

$$\mathcal{L}_t(v, \mu) = J_t(v) - \int_0^1 \mu v\, dx_2^t \quad , \quad t \geq 0$$

the Lagrangian defined on $H^1(\Omega_t(\alpha)) \times L^2_+(\Gamma_{2t}(\alpha))$. If the solution of $(\mathcal{P}(\alpha))_t$ is sufficiently regular, then $\mathcal{L}_t$ has a unique saddle point (u_t^*, μ_t^*) on $\mathcal{U}_t(\alpha)) \times L^2_+(\Gamma_{2t}(\alpha))$. Moreover

$$u_t^* = u_t(\alpha), \mu_t^* = \frac{\partial u_t^*}{\partial n} \sqrt{1 + \left[\alpha' + t\left(\frac{\partial V_1}{\partial x_1}\alpha' + \frac{\partial V_1}{\partial x_2}\right)\right]^2}.$$

It is well–known that

$$(2.1)\qquad \begin{aligned} E_t(\alpha) = J_t\left(u_t(\alpha)\right) &= \min_{v \in \mathcal{U}_t(\alpha)} \sup_{\mu \in L^2_+(\Gamma_{2t}(\alpha))} \mathcal{L}_t(v, \mu) = \\ &= \max_{\mu \in L^2_+(\Gamma_{2t}(\alpha))} \inf_{v \in \mathcal{U}_t(\alpha))} \mathcal{L}_t(v, \mu). \end{aligned}$$

Functions of min max type can be quite easily differentiated. It is known (see [2]) that

$$\dot{E}(\alpha) = \frac{d}{dt} E_t(\alpha)\Big|_{t=0} = \mathcal{L}_{,t}\left(u_0^*(\alpha), \mu_0^*\right)\Big|_{t=0},$$

where $\mathcal{L}_{,t}$ denotes the partial derivative of $\mathcal{L}_t$ with respect to the real parameter t. A direct calculation yields:

$$\begin{aligned} &\mathcal{L}_{,t}\left(u_0^*(\alpha), \mu_0^*\right)\Big|_{t=0} = \\ &= \frac{1}{2}\int_{\partial\Omega(\alpha)} |\nabla u_0^*(\alpha)|^2 (V \cdot n)\, ds - \int_{\partial\Omega(\alpha)} f u_0^*(\alpha)(V \cdot n)\, ds - \int_0^1 \nabla\left(u_0^*(\alpha)\mu_0^*\right) \cdot V\, dx_2 = \\ &= \frac{1}{2}\int_{\Gamma_2(\alpha)} |\nabla u_0^*(\alpha)|^2 V_1 n_1\, ds - \int_{\Gamma_2(\alpha)} f u_0^*(\alpha) V_1 n_1\, ds - \int_0^1 \nabla\left(u_0^*(\alpha)\mu_0^*\right) \cdot V\, dx_2, \end{aligned}$$

using formulas of differentiation of integrals where the integrand as well as the domain of integration depend on a parameter t (see [1], [3]), as obvious.

Now, let $I(\alpha)$ be an interval, where the contact is realized, i.e.

$$I(\alpha) \subset \{x \in [0,1] \mid u_0^*(\alpha)(x) = 0\}$$

and let V be a vector field such that supp $V_1(\alpha(x_2), x_2) \subset I(\alpha)$. Then for such a vector field one has

$$\dot{E}(\alpha) = \frac{1}{2}\int_{I(\alpha)} \left|\frac{\partial u_0^*}{\partial n}\right|^2 V_1 n_1 \sqrt{1+(\alpha')^2}\,dx_2 - \int_{I(\alpha)} (\nabla u_0^*(\alpha)\cdot V)\,\mu_0^*\,dx_2, \tag{2.2}$$

as $(\nabla \mu_0^*(\alpha)\cdot V)u_0^*(\alpha) = 0$ on $I(\alpha)$ $(u_0^*(\alpha) \equiv 0$ on $I(\alpha))$. From the same reason one has

$$(\nabla u_0^*(\alpha)\cdot V)\,\mu_0^* = (\nabla u_0^*\cdot n)\,(V\cdot n)\,\mu_0^* = \frac{\partial u_0^*}{\partial n}\mu_0^* V_1 n_1.$$

From this, (2.1) and (2.2) we finally obtain

$$\begin{aligned}\dot{E}(\alpha) &= \frac{1}{2}\int_{I(\alpha)} \left|\frac{\partial u_0^*}{\partial n}\right|^2 V_1 n_1 \sqrt{1+(\alpha')^2}\,dx_2 - \int_{I(\alpha)} \left|\frac{\partial u_0^*}{\partial n}\right|^2 V_1 n_1 \sqrt{1+(\alpha')^2}\,dx_2 = \\ &= -\frac{1}{2}\int_{I(\alpha)} \left|\frac{\partial u_0^*}{\partial n}\right|^2 V_1 n_1 \sqrt{1+(\alpha')^2}\,dx_2\end{aligned} \tag{2.3}$$

for any vector field $V = (V_1, 0)$, supp $V_1(\alpha(x_2), x_2) \subset I(\alpha)$.

We say that $\Gamma_2(\alpha^*)$ is an internal solution of (P) if α^* solves (P) and

$$\begin{cases} C_0 < \alpha^*(x_2) < C_1 & \forall x_2 \in [0,1], \\ |\alpha^*(x_2) - \alpha^*(\bar{x}_2)| \le \bar{C}_2 |x_2 - \bar{x}_2| & \forall x_2, \bar{x}_2 \in [0,1] \end{cases} \tag{2.4}$$

for some constant $0 < \bar{C}_2 < C_2$.

Let α^* satisfy (2.4). Then it is readily seen that α_t^*, graph of which is $\Gamma_{2t}(\alpha^*)$, satisfies (2.4) as well, provided $t > 0$ sufficiently small. In order to remove the constant constraint volume meas $\Omega_t(\alpha) = C_3$, we introduce the following Lagrangian:

$$L_t(\alpha) = E(t) + \lambda\,(\text{meas}\,\Omega_t(\alpha) - C_3)\,, \quad \lambda \in \mathbf{R}^1, \quad t > 0.$$

Let $I(\alpha^*)$ be a contact set for an optimal solution $u_0(\alpha^*)$ and let α^* satisfy (2.4). By $\bar{V} = (\bar{V}_1, 0)$ we denote any vector field such that supp $\bar{V}_1(\alpha(x_2), x_2) \subset I(\alpha^*)$. Then necessary condition for α^* to be an optimal solution is the existence of $\lambda^* \in \mathbf{R}^1$ such that

$$\dot{L}(\alpha^*) = \frac{d}{dt}L_t(\alpha^*)\Big|_{t=0} = \dot{E}(\alpha^*) + \lambda^*\frac{d}{dt}(\text{meas}\,\Omega_t(\alpha^*))\Big|_{t=0} = 0$$

i.e.

$$\frac{1}{2}\int_{I(\alpha^*)} \left|\frac{\partial u_0^*}{\partial n}\right|^2 \bar{V}_1 n_1 \sqrt{1+(\alpha'^*)^2}\,dx_2 = \lambda^* \int_{I(\alpha^*)} \bar{V}_1 n_1 \sqrt{1+(\alpha'^*)^2}\,dx_2$$

for any vector field $\bar{V}$ with about mentioned property. From this $|\frac{\partial u_0^*}{\partial n}| =$ const. on $I(\alpha^*)$ and as $\frac{\partial u_0^*}{\partial n}$ is non–negative on $\Gamma_2(\alpha^*)$, we finally obtain

$$\frac{\partial u_0^*}{\partial n} = \text{const.} \geq 0 \text{ on } I(\alpha^*).$$

References

[1] J. Haslinger and P. Neittaanmäki, "Finite element approximation for shape optimal design: Theory and applications", J. Wiley & Sons, Chichester, 1988.

[2] R. Correa and A. Seeger, *Directional derivatives in minimax problems*, Numer. Funct. Anal. and Optimiz. **7**, (1984), p. 145–156.

[3] E.J. Haug, K.K. Choi and V. Komkov, *Design sensitivity analysis of structural systems*, Mathematics in Science and Engineering **77**, Academic Press, 1986.

[4] R.L. Benedict and J.E. Taylor, *Optimal design for elastic bodies in contact*, in "Optimization of distributed parameters structures", Part I & II, Nato Advances Study Institute Series, Series E, Sijthoff & Noordhoff, Alphen aan den Rijn, 1981.

[5] I. Hlaváček, J. Haslinger, J. Nečas and J. Lovíšek, *Numerical solution of variational inequalities in mechanics*, Springer series in Applied Mathematical Sciences **66**, Springer–Verlag, 1988.

DYNAMICS OF RIGID BODIES WITH DRY FRICTION AND PARTIALLY ELASTIC COLLISIONS.

Michel JEAN
CNRS, Montpellier, France.

ABSTRACT.

This paper deals with dynamical problems in the presence of unilateral contact and dry friction. Possible shocks may be either inelastic or partially elastic. The application in view is the motion of a collection of balls enclosed between walls and coming into contact with each other. A double time scale argument is evoked to justify the law governing shocks. It is observed that constraints are *kinematically dependent*. A numerical method is proposed and some computed examples are produced. Since the method proves efficient, one expects to deal in the future with a large number of balls, using a super computer, so as to generate the model of an aggregate. Everything in the sequel is presented in the *two-dimensional* setting.

1. UNILATERAL CONTACT CONDITION AND DRY FRICTION LAW.

First, for simplicity, the case of a single particle P lying at some time t in the vicinity of a possibly moving obstacle O is considered. A *local frame* is defined by its origin P', considered as a particle of O , by the unit normal vector N directed towards the free region, and the unit tangent vector T. As a particular choice, we shall assume P' to equal the particle of O coinciding with the orthogonal projection of P at time t on the boundary of O (Fig. 1). One defines:

q_N normal coordinate of P in the local frame.

$\mathcal{R} = (\mathcal{R}_T , \mathcal{R}_N)$ components of the reaction in the local frame.

$\mathcal{U} = (\mathcal{U}_T , \mathcal{U}_N)$ components of the relative velocity in the local frame.

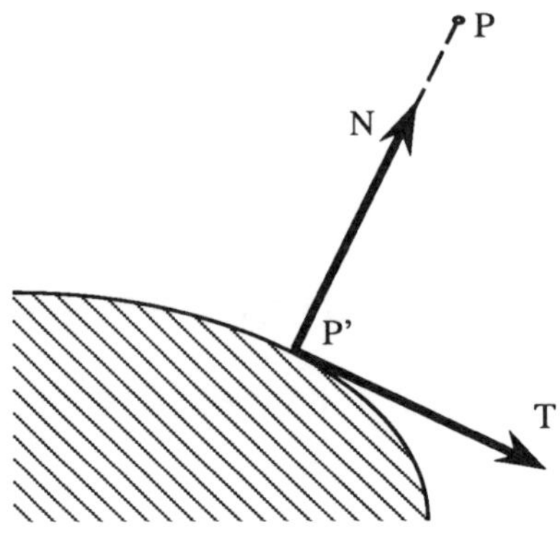

Fig. 1

Unilateral conditions: They are expressed as follows:
Impenetrability: $q_N \geq 0$.
When the particle is in contact, the obstacle is supposed not to attract the particle: $q_N = 0 \Rightarrow \mathcal{R}_N \geq 0$.
When the particle is not in contact, the reaction vanishes: $q_N > 0 \Rightarrow \mathcal{R}_N = 0$.

These relations may be assembled in the usual form of a set of *complementarity conditions*

(1.1) $$q_N \geq 0 \ , \ \mathcal{R}_N \geq 0 \ , \ q_N \mathcal{R}_N = 0,$$

or equivalently

(1.2) $$\mathcal{R}_N \geq 0 \text{ and } \forall \mathcal{S}_N \geq 0 \ (\mathcal{S}_N - \mathcal{R}_N)\, q_N \geq 0 \,,$$

also equivalent to

(1.3) $$q_N \geq 0 \text{ and } \forall p_N \geq 0 \ (p_N - q_N)\, \mathcal{R}_N \geq 0 \,.$$

The graph of relation (1.1) is represented on Fig. 2.

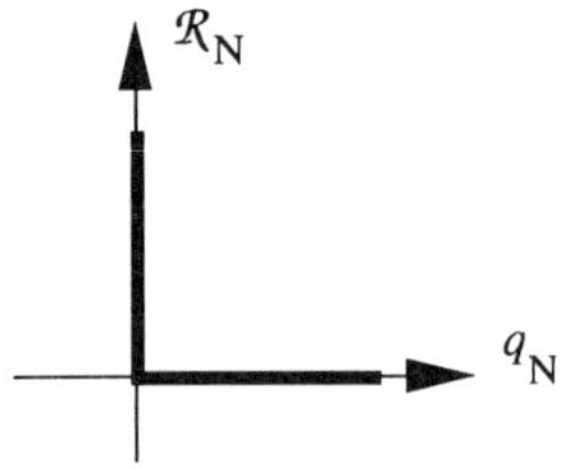

Fig. 2

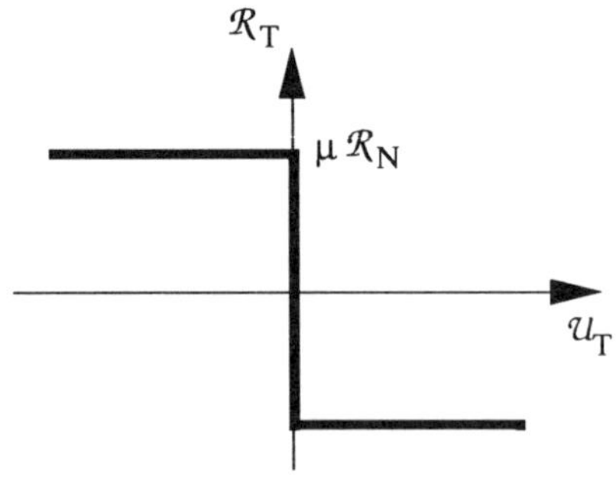

Fig. 3

Shock law: In dynamical problems, the presence of unilateral constraints involves the possibility of *collisions*. A shock law thus has to be prescribed.

Velocities will be assumed to be functions of time with *bounded variation*. This allows one to define, at every instant, the right and left velocities $\mathcal{U}^+$ and $\mathcal{U}^-$ of the concerned particle; these two limits are expected to be different if a collision occurs. In this case, the mutual action of the colliding bodies cannot be described in terms of reaction forces, but in terms of reaction *impulses*, i.e. vector measures on the considered time interval. Recall that any finite-dimensional vector measure may be represented by a *density function* relative to a suitably chosen nonnegative real measure called a *base measure*. So the vector function of time denoted by $\mathcal{R}$ in the above statements will more generally be understood as the density function of the reaction impulse with regard to some base measure .Throughout an interval of smooth motion, the base measure may be taken equal to the Lebesgue measure so $\mathcal{R}$ properly represents a force. In contrast, at the instant t_c of a collision, the base measure is expected to exhibit an *atom* which may be viewed as the Dirac mass at point t_c ; then $\mathcal{R}(t_c)$ equals the value of the reaction impulse. Anyway, because relations (1.1) to (1.3) are positively homogeneous with regard to $\mathcal{R}$, all choices of base measures relative to which the reaction impulse admits a density functions generate equivalent mechanical statements (J. J. Moreau [1] [4])

The inelastic shock law. The assumption of *standard inelastic shock,* proposed by J. J. Moreau, may equivalently be stated as follows (M. Jean [2])

(1.4) $$q_N = 0 \Rightarrow \mathcal{U}^+_N \geq 0 .$$

(1.5) $$q_N = 0 \text{ and } \mathcal{R}_N > 0 \Rightarrow \mathcal{U}^+_N = 0 .$$

The concept of a shock properly pertains to problems of Dynamics. However, when deriving algorithms of time-discretization, the formulation (1.5) has also been successfully applied to the treatment of *quasistatic* problems, i.e. evolutions in which inertia terms are neglected; the resulting algorithms are similar to those based on (1.1).

Another shock law. Here is a classical alternative assumption:

(1.6) $$q_N = 0 \Rightarrow \mathcal{U}^+_N = - e\, \mathcal{U}^-_N .$$

where $e \in [0,1]$ is called the *restitution coefficient*. The case e=0 corresponds to inelastic shock In the next section, the existence of e will be justified to a certain extent, by a method of double scaling relative to time. To this end, the colliding bodies will be viewed as locally slightly deformable, with viscoelastic behaviour. For another example of use of this shock law, see [3].

Friction laws. *Coulomb's law* is considered here for its interest in many applications. It relates $\mathcal{R}$ to the sliding velocity $\mathcal{U}_T$. In the two-dimensional case, when the particle is in contact:

$$\mathcal{R}_T \in [-\mu\, \mathcal{R}_N , \mu\, \mathcal{R}_N] \;, \quad \mu \textit{ friction coefficient},$$
$$\mathcal{U}_T > 0 \;\Rightarrow\; \mathcal{R}_T = -\mu\, \mathcal{R}_N ,$$
$$\mathcal{U}_T < 0 \;\Rightarrow\; \mathcal{R}_T = \mu\, \mathcal{R}_N .$$

Fig. 3 shows the graph of this relation. J. J. Moreau [4] has proposed the equivalent writing

(1.7) $$\mathcal{R}_T \in \partial\Psi^*_C(-\mathcal{U}_T),$$

where C is the interval $[-\mu\, \mathcal{R}_N , \mu\, \mathcal{R}_N]$. Classically, Ψ_C denotes the *indicator function* of this subset of the real line, Ψ^*_C the Fenchel conjugate of this function, i.e. the *support function* of C and $\partial\Psi^*_C$ the *subdifferential* of Ψ^*_C. Another equivalent statement is the "principle of maximal dissipation ", namely

(1.8) $$\mathcal{R}_T \in C \text{ and } \forall\, \mathcal{S}_T \in C \quad (\mathcal{S}_T - \mathcal{R}_T)\, \mathcal{U}_T \geq 0.$$

The above also applies to the three-dimensional case, the interval being replaced by a disk.

The same formalism may be extended to the analysis of a shock, provided that $\mathcal{R}$ is understood as the density of the reaction impulse with respect to a base measure. Since $\mathcal{U}$ is then discontinuous, one has to specify what to take as $\mathcal{U}_T$ in the foregoing statements. *We decide to use $\mathcal{U}^+_T$ as the concerned value of the velocity.* This choice amounts to combine frictional effects with the assumption of inelastic collision and may also be justified by an argument of double scaling in time .

Kinematical relations. The position of the particle P being specified by its absolute coordinates $q=(q_1, q_2)$, there exists an affine relation between the value $\dot{q}$ of the derivative dq/dt for an arbitrary motion passing through this position and the corresponding velocity $\mathcal{U}$ of P in the local frame, say

(1.9) $$\mathcal{U} = H^*(q)\, \dot{q} + \mathcal{U}_0(q) .$$

The known vector $\mathcal{U}_0(q)$ is the negative of the boundary velocity; it vanishes in the case of fixed obstacle. Rather than on the 2×2 matrix H*(q), emphasis will be placed in the sequel on its transpose H(q).The latter is used to express the absolute components $R=(R_1, R_2)$ of the reaction force from its components $\mathcal{R}=(\mathcal{R}_T, \mathcal{R}_N)$ in the local frame, namely

(1.10) $$R = H(q)\, \mathcal{R}.$$

Generalization to a system of particles or rigid bodies. Extending the above writing is straightforward (M. Jean and J. J. Moreau [5]). The equations governing contact and friction have to be written for every particle P^α candidate for contact, $\alpha = 1,...,\chi$. Henceforth, we shall denote by $\mathcal{U}_T = (\mathcal{U}_T^{\,1},...,\mathcal{U}_T^{\,\alpha},...)$ the χ-dimensional vector consisting of the tangential components of the respective local velocities $\mathcal{U}^\alpha$. Similarly, $\mathcal{U}_N = (\mathcal{U}_N^{\,1},...,\mathcal{U}_N^{\,\alpha},...)$ consists of the normal components of these velocities, while $\mathcal{R}_T = (\mathcal{R}_T^{\,1},...,\mathcal{R}_T^{\,\alpha},...)$ and $\mathcal{R}_N = (\mathcal{R}_N^{\,1},...,\mathcal{R}_N^{\,\alpha},...)$ represent in the same way the corresponding reactions. Also $q_N = (q_N^{\,1},...,q_N^{\,\alpha},...)$ denotes the χ-

dimensional vector made of the normal coordinates of the particles with regard to the respective local frames. By writing relations (1.2) and (1.8) for each particle P^α, then using the standard scalar product and the standard ordering of $\mathbf{R}^\chi$, one obtains

(1.11) $$\mathcal{R}_N \geq 0 \text{ and } \forall\, \mathcal{S}_N \geq 0 \quad (\mathcal{S}_N - \mathcal{R}_N).q_N \geq 0\,,$$

(1.12) $$\mathcal{R}_T \in \mathcal{C} \text{ and } \forall\, \mathcal{S}_T \in \mathcal{C} \quad (\mathcal{S}_T - \mathcal{R}_T).\mathcal{U}^+{}_T \geq 0\,.$$

Here $\mathcal{C}$ denotes the following subset of $\mathbf{R}^\chi$

$$\mathcal{C} = \{\mathcal{R}_T=(\mathcal{R}_T{}^1,\ldots,\mathcal{R}_T{}^\alpha,\ldots) \mid \forall\alpha\,,\ \mathcal{R}_T{}^\alpha \in \mathcal{R}_N{}^\alpha\,\mathcal{D}\,\},$$

with $\mathcal{D}$ equal to the interval $[-\mu, \mu]$. This definition of $\mathcal{C}$ may be condensed into the writing $\mathcal{C}=\mathcal{R}_N\,\mathcal{D}$.

Let a position of the system be specified by $q=(q_1, q_2,\ldots,q_n)$. Relations between $\dot{q}$ and the relative velocities at the various contact points may be deduced from kinematical inspection, namely

$$\mathcal{U} = H^*(q)\,\dot{q} + \mathcal{U}_0(q)\,.$$

Here $H^*(q)$ is a linear mapping of $\mathbf{R}^n$ to $\mathbf{R}^{2\chi}$. Its transpose $H(q)$ serves to express the representative $R=(R_1, R_2,\ldots,R_n)$, in the sense of Analytical Dynamics, of the set of the various contact reactions, the respective local components of which constitute the 2χ-dimensional vector $\mathcal{R}$. The known term $\mathcal{U}_0(q)$ vanishes in the case of time-independent constraints (contact with some fixed external obstacle or contact between bodies which both are part of the investigated system).

In some usual cases, $H(q)$ is *injective*, i.e. the element R of $\mathbf{R}^n$ vanishes only if all local reactions vanish. Equivalently $H^*(q)$ is *surjective*, which means that for any chosen set of values for the local velocities $\mathcal{U}$, there exists a value of $\dot{q} \in \mathbf{R}^n$ allowing to meet this choice. In contrast, when *constraints are kinematically dependent* , $H(q)$ is no more injective. A typical example of such a situation is provided by two particles at the ends of a rigid rod. Also, in the case we shall later present, of a collection of balls in contact with each other, one cannot expect H to be injective.

The problem equations. Summing up, the motion of a system with finite degree of freedom in the presence of unilateral contact and dry friction is governed by:

* The equation of dynamics,

$$M(q)\,\ddot{q} = -F(t, q, \dot{q}\,) + R\,,$$

where $M(q)$ is the inertia matrix. The known function F involves the $\dot{q}$-dependent terms of the Lagrange equations and, possibly, some explicitly given internal or external forces. In general, this equation has to be understood in the sense of measures (J. J. Moreau [9]) and is written more precisely in the form

$$M(q)\, d\dot{q} = -F(t, q, \dot{q})\, dt + R\, ds.$$

In accordance with the notations previously used, ds is a nonnegative base measure and the $\mathbf{R}^n$-valued measure R ds constitutes the representative, in the sense of Analytical Dynamics, of the set of the reaction impulses at the various contact points.

* The initial conditions.
* The dual relations. $\mathcal{U} = H^*(q)\,\dot{q} + \mathcal{U}_0(q)$ and $R = H(q)\,\mathcal{R}$.
* The unilaterality conditions.
* The shock law.
* Coulomb's law.

The question of the existence and uniqueness of the solution to this problem does not lie within the scope of this paper.

2. THE RESTITUTION COEFFICIENT DISCUSSED THROUGH DOUBLE SCALING IN TIME.

In this section the obstacle O will be assumed fixed and conceived in a way different from the foregoing. The reaction force experienced by a particle P, candidate for contact, is interpreted as arising from a *viscoelastic body* admitting O as unstrained reference configuration. Its normal component is assumed to obey

(2.1) $$\mathcal{R}_N = -f(q_N, \dot{q}_N) \qquad \mathcal{R}_N \geq 0 \qquad \mathcal{R}_N\, q_N \leq 0,$$

without any sign condition imposed upon q_N. With regard to the couple of variables q_N, $\dot{q}_N$, the function f is assumed to be Lipschitz-continuous, non decreasing and vanishing at (0,0). For instance

(2.1') $$\mathcal{R}_N = -k\, q_N - v\, \dot{q}_N \qquad k \geq 0 \qquad v \geq 0.$$

In addition, one supposes that the tangential component of the reaction force satisfies Coulomb's law with sliding velocity defined relatively to O . By putting $C = [-\mu\, \mathcal{R}_N, \mu\, \mathcal{R}_N]$,

(2.2) $$\mathcal{R}_T \in C \quad \text{and} \quad \forall\, \mathcal{S}_T \in C \quad (\mathcal{S}_T - \mathcal{R}_T)\, \mathcal{U}_T \geq 0.$$

For simplicity, the sequel will be restricted to the case where a single particle of the system, distinguished by α, may come into contact . As in Section 1, the equations of motion

$$M(q)\,\ddot{q} = -F(t, q, \dot{q}) + R,$$
$$\mathcal{U} = H^*(q)\,\dot{q} + \mathcal{U}_0(q),$$
$$R = H(q)\,\mathcal{R},$$

yield the following, involving the local variables associated with the considered particle,

$$(2.3) \qquad \mathcal{M}^{\alpha}(q)\ \dot{\mathcal{U}}^{\alpha} = -\ \mathcal{F}^{\alpha}(t, q, \dot{q}) + \mathcal{R}^{\alpha}.$$

By $\mathcal{M}^{\alpha}$ is denoted the 2×2 matrix $\mathcal{M}^{\alpha}(q)=[H^{\alpha*}(q)M(q)^{-1}H^{\alpha}(q)]^{-1}$, called the *equivalent inertia matrix*. As before, $H^{\alpha}(q)$ is assumed to be injective, so that $\mathcal{M}^{\alpha}$ is symmetric positive definite. In addition

$$\mathcal{F}^{\alpha}(t, q, \dot{q}) = \mathcal{M}^{\alpha}(q)\ [H^{\alpha*}(q)M(q)^{-1}F(t, q, \dot{q}) - d\mathcal{U}_0(q)/dt - dH^{\alpha*}(q)/dt\ \dot{q}]\ .$$

Kinematics yields

$$\mathcal{U}^{\alpha}_N = \dot{q}^{\alpha}_N.$$

One intends to study a smooth motion with initial conditions

$$q^{\alpha}_N(0) = 0\ , \qquad \dot{q}^{\alpha}_N(0) = \mathcal{U}^{\alpha}_N(0) < 0\ .$$

For brevity, subscripts and superscripts α will henceforth be omitted. One puts

$$M = \mathcal{M}(q(0))\ , \qquad M=\begin{bmatrix} M_{NN} & M_{NT} \\ M_{TN} & M_{TT} \end{bmatrix},$$

$$m_N = M_{NN} - M_{NT}M^{-1}_{TT}M_{TN},$$

$$m_T = M_{TT} - M_{TN}M^{-1}_{NN}M_{NT}\ .$$

Since M is positive definite, $m_N > 0$ and $m_T > 0$.

Assume that the interaction between the particle and the obstacle takes place in a short interval of time, say [0,t']. The technique of *double scaling in time* rests on the assumption that t' is sufficiently small for the functions $t \rightarrow \mathcal{M}(q(t))$ and $t \rightarrow \mathcal{F}(t,q(t), \dot{q}(t))$ to be nearly constant on [0,t']. This allows one to use the equations of Dynamics in an approximate form

$$(2.3') \qquad M\dot{\mathcal{U}} = -\ F + \mathcal{R}.$$

First case:

$$\mathcal{U}_T(0) > 0, \qquad 1 + \mu\, M_{NT}M^{-1}_{TT} > 0, \quad -\mu < M_{TN}M^{-1}_{NN} < \mu.$$

One defines $m^{+}_N = m_N/(1 + \mu\, M_{NT}M^{-1}_{TT})$. The solution to (2.3') joined with (2.1') and (2.2) is found to be

$$\mathcal{U}_T(t') = \max\{0\ , \mathcal{U}_T(0) - (\mu + M_{TN}M^{-1}_{NN})m^{+}_N/m_T(\mathcal{U}_N(t') - \mathcal{U}_N(0))\}$$

Second case:

$$\mathcal{U}_T(0) < 0, \qquad 1 - \mu\, M_{NT}M^{-1}_{TT} > 0, \quad -\mu < M_{TN}M^{-1}_{NN} < \mu$$

One defines $m^{-}_N = m_N/(1 - \mu\, M_{NT}M^{-1}_{TT})$ and similar discussion yields

$$\mathcal{U}_T(t') = \min\{0\,,\, \mathcal{U}_T(0) + (\mu - M_{TN}M_{NN}^{-1})m_N^-/m_T\,(\mathcal{U}_N(t') - \mathcal{U}_N(0))\}$$

Third case:

$$\mathcal{U}_T(0) = 0$$

The solution is found to be $\mathcal{U}_T(t') = 0$.

Description of the limit process in terms of measures. Solve

$$M\, d\dot{q} = \mathcal{R}\, ds\,,$$

$$\mathcal{R}_N \geq 0 \quad \text{and} \quad \forall\, \mathcal{S}_N \geq 0 \quad (\mathcal{S}_N - \mathcal{R}_N)\, q_N \geq 0\,,$$

$$\mathcal{R}_T \in \mathcal{R}_N\, \mathcal{D} \quad \text{and} \quad \forall\, \mathcal{S}_T \in \mathcal{R}_N \mathcal{D} \quad (\mathcal{S}_T - \mathcal{R}_T)\, \mathcal{U}_T \geq 0\,.$$

Here, the base measure ds equals the Dirac measure at time 0. The values taken at this point by the density functions $\mathcal{R}_T$ and $\mathcal{R}_N$ simply equal the corresponding impulsive quantities. After integrating both members of the measure differential equation on the singleton {0} and separating the components, one obtains

$$m_N(\mathcal{U}_N^+ - \mathcal{U}_N^-) = \mathcal{R}_N - M_{NT}M_{TT}^{-1}\, \mathcal{R}_T\,,$$

$$m_T\,(\mathcal{U}_T^+ - \mathcal{U}_T^-) = \mathcal{R}_T - M_{TN}M_{NN}^{-1}\, \mathcal{R}_N\,.$$

Assume:

$$1 + \mu\, M_{NT}M_{TT}^{-1} > 0, \quad 1 - \mu\, M_{NT}M_{TT}^{-1} > 0, \quad -\mu < M_{TN}M_{NN}^{-1} < \mu \text{ and } \mathcal{U}_N^+ - \mathcal{U}_N^- \geq 0.$$

In the first of the above cases, one supposes $\mathcal{U}_T^- > 0$; then, the solution is found to be

$$\mathcal{U}_T^+ = \max\{0\,,\, \mathcal{U}_T^- + (\mu - M_{TN}M_{NN}^{-1})m_N^+/m_T(\mathcal{U}_N^+ - \mathcal{U}_N^-)\}$$

In the second case one supposes $\mathcal{U}_T^- < 0$; then the solution is found to be

$$\mathcal{U}_T^+ = \max\{0\,,\, \mathcal{U}_T^- - (\mu + M_{TN}M_{NN}^{-1})m_N^-/m_T((\mathcal{U}_N^+ - \mathcal{U}_N^-))\}$$

In the third case one supposes $\mathcal{U}_T^- = 0$; then $\mathcal{U}_T^+ = 0$.

Conclusion. Provided that

$$\|M_{NT}M_{TT}^{-1}\| < 1/\mu\,, \quad \|M_{TN}M_{NN}^{-1}\| < \mu\,, \quad \mathcal{U}_N^+ - \mathcal{U}_N^- \geq 0\,,$$

it appears that the right and left sliding velocities, solutions in the sense of measures to the contact/friction equation, are given by the same formula than the respective values, at the beginning and the end of contact, of the sliding velocities corresponding to the smooth process. However, all possible information has not been drawn here from the above treatment of the collision as smooth process. Another expression relating $\mathcal{U}(t')$, $\mathcal{U}(0)$, $\mathcal{R}_T$, $\mathcal{R}_N$, allows one to derive $\mathcal{U}_N(t') - \mathcal{U}_N(0)$. This, unfortunately, does not yield a simple shock law except in the two following circumstances:

i) The process is assumed to dissipate just enough energy for $\mathcal{U}_N^+ = 0$ to hold. This is an *inelastic shock*.

ii) The matrix M is supposed to be diagonal: $M_{NT} = 0$, $M_{TN} = 0$. Then $\mathcal{U}_N(t')$ depends only upon $\mathcal{U}_N(0)$. As an example, when

$$\mathcal{R}_N = -f(q_N, \dot{q}_N) = -k\, q_N - v\, \dot{q}_N ,$$

a linear differential equation is involved. With v small enough one finds

$$\alpha = -\pi v/\sqrt{m_N k}\,, \qquad \omega^2 = 4\pi^2 - \alpha^2 .$$

The contact duration is approximately π/ω and

$$\mathcal{U}_N(t') = -e^{\alpha/2}\, \mathcal{U}_N(0) .$$

Thus the coefficient of restitution approximately equals $e^{\alpha/2}$.

In other circumstances the argument of double scaling in time cannot be brought to a definite conclusion. It has been mentionned by R. M. Brach, [6], that a wild use of the concept of coefficient of restitution may lead to violation of energy balance. K. L. Johnson [8] has discussed the case of large impact velocity so that *plastic deformation* of the material must be taken into account.

Observe that for a single mass-point the equivalent inertia matrix is diagonal. Such is also the case for a disk:

$$M_{TT} = (m^{-1} + I^{-1} r^2)^{-1}, \qquad M_{NN} = m ,$$

where m denotes the mass of the disk, I its moment of inertia with respect to the center, and r its radius.

3. TIME DISCRETIZATION IN THE PRESENCE OF UNILATERAL CONTACT AND DRY FRICTION.

The dynamical equation. For simplicity, the equation governing the system is supposed to be linear with constant coefficients, namely in the case where the function $t \to q(t) \in \mathbf{R}^n$ is smooth enough

$$M\ddot{q} + V\dot{q} + Kq = P + R .$$

Here the $\mathbf{R}^n \times \mathbf{R}^n$ matrices M, V, K respectively represent inertia, viscosity and stiffness; P is an explicitly given external loading while R represents the set of the reactions of unilateral constraints.

More generally, in order to treat possible collisions on the same footing as smooth motions, one makes provision for the function $t \to q(t)$ to be discontinuous, with bounded variation. The above differential equation is thus replaced, as in Section 1, by the *measure differential equation*

$$M\, d\dot{q} + V\, dq + Kq\, dt = P\, dt + R\, ds ,$$

where the function $t \to R(t) \in \mathbf{R}^n$ now represents the density of the reaction impulse with respect to some base measure ds. Integrating both members over a time interval $]t_i, t_{i+1}]$ yields

$$M(\dot{q}(i+1) - \dot{q}(i)) + V(q(i+1) - q(i)) + K \int q\, dt = \int P\, dt + \int R\, ds.$$

After choosing a constant $\theta \in]0,1]$ and putting $h = t_{i+1} - t_i$, one uses the following approximants of the respective terms:

$$q(i+1)-q(i)= ((1-\theta)\, \dot{q}(i) + \theta\, \dot{q}(i+1))h ,$$

$$\int q\, dt = [(1-\theta)q(i)+\theta q(i+1)]h ,$$

$$\int P\, dt = [(1-\theta)P(i)+\theta P(i+1)]h ,$$

$$\int R ds= R(i+1)h .$$

For the ease of computation, one introduces

$$\hat{q}(i) = q(i) + (1-\theta)\, \dot{q}(i)h,$$

$$W = \hat{M}^{-1} ,$$

$$\hat{M} = M + \theta hV + \theta^2 h^2 K ,$$

$$\hat{V} = V + (2\theta-1)hK ,$$

$$A(i)h = [-\hat{V}\, \dot{q}(i) \;\; -K\, \hat{q}(i) + (1-\theta)P(i)+\theta P(i+1)]h ,$$

$$\dot{q}(i+1) = \dot{q}(i) + WA(i)h + WR(i+1)h , \tag{3.1}$$

This procedure may be viewed as an adaptation of the so-called *theta method* enabling it to treat the case where the function $\dot{q}$ is discontinuous. Moreover, by making θ=1, V=0, M=0, one obtains a finite difference scheme suitable to compute quasi-static evolutions.

The value $\hat{q}(i)$ defines an approximated configuration of the system at time t_i , and $\hat{q}(i+1)$ an approximated configuration at time t_{i+1}. The choice of these expressions is suggested by the approximate relation (3.4) below. It allows one to deal with both the unilateral conditions and the law of a possible shock

The linear mappings H and H* introduced in Section 1 will be used to connect $\mathcal{U}(i+1)$ (an approximant of $\mathcal{U}^+$ at time t_{i+1}), $\mathcal{R}(i+1)$ (an approximant of $\mathcal{R}$ at this time) with R(i+1) and $\dot{q}(i+1)$

(3.2) $$\mathcal{U}(i+1) = H^* \dot{q}(i+1) + \mathcal{U}_0(i),$$

(3.3) $$R(i+1) = H\, \mathcal{R}(i+1).$$

The values of H and H* in these relations will be calculated at some estimated position $q(t_{i+1})$. Apart from this estimation of $q(t_{i+1})$ the whole discretization scheme appears to be of the *implicit* type. Its *stability* may be ascribed to this feature.

When the distance between the positions of a contacting particle at two successive time steps is small enough for the rotation of the corresponding normal vector to be neglected, the following approximate relation may be used

(3.4) $$q_N(i+1) - q_N(i) = h\, \mathcal{U}_N(i+1),$$

where $q_N(i)$ and $q_N(i+1)$ denote approximants of $q_N(t_i)$ and $q_N(t_{i+1})$.

The case of an inelastic shock. The conditions (1.2) of unilateral contact are written for the approximate values of $\mathcal{R}_N$ and q_N

(3.5) $$\mathcal{R}_N(i+1) \geq 0 \quad \text{and} \quad \forall\ \mathcal{S}_N \geq 0 \quad (\mathcal{S}_N - \mathcal{R}_N(i+1)).q_N(i+1) \geq 0.$$

Relation (3.4), together with $q_N(i) = 0$, yields the implication

$$q_N(i+1) = 0 \ \Rightarrow\ \mathcal{U}_N(i+1) = 0.$$

So, under (3.4), the law of inelastic shock is satisfied as soon as the conditions of unilateral contact hold.

Furthermore, Coulomb's law may be written down with the same notations as in (1.12)

(3.6) $$\mathcal{R}_T(i+1) \in C \quad \text{and} \quad \forall\ \mathcal{S}_T \in C \quad (\mathcal{S}_T - \mathcal{R}_T(i+1))\ \mathcal{U}_T(i+1) \geq 0,$$

$$C = \mathcal{R}_N(i+1)\ \mathcal{D}.$$

Eliminating $\dot{q}(i+1)$ and R(i+1) between equations (3.1) to (3.3), one obtains a relation between contact velocities and contact reactions,

(3.7) $$\mathcal{U}(i+1) = H^*WHh\mathcal{R}(i+1) + E_0,$$

where

$$E_0 = \mathcal{U}_0(i) + H^*(\dot{q}(i) + WhA(i)).$$

Replacing $\mathcal{U}_T(i+1)$ in the conditions of unilateral contact (3.5) by the value drawn from (3.7), and replacing $q_N(i+1)$ in the friction law (3.6) by the value drawn from (3.4) and (3.7), one gets a system of *two coupled variational inequalities* , the unknown of which are $\mathcal{R}_T(i+1)$, $\mathcal{R}_N(i+1)$ or, more precisely, the impulses $h\mathcal{R}_T(i+1)$, $h\mathcal{R}_N(i+1)$,

(3.8) $\qquad h\mathcal{R}_N \geq 0$ and $\forall\, \mathcal{S}_N \geq 0 \quad (\mathcal{S}_N - h\mathcal{R}_N).(W_{NN}\, h\mathcal{R}_N + W_{NT}\, h\mathcal{R}_T + E'_{0N}) \geq 0$

(3.9) $\qquad h\mathcal{R}_T \in \mathcal{C}$ and $\forall\, \mathcal{S}_T \in \mathcal{C} \quad (\mathcal{S}_T - h\mathcal{R}_T).(W_{TN}\, h\mathcal{R}_N + W_{TT}\, h\mathcal{R}_T + E'_{0T}) \geq 0$,

$\qquad \mathcal{C} = h\mathcal{R}_N \mathcal{D}$,

(the subscript i+1 is omitted)

$$E'_{0T} = E_{0T} \quad , \; E'_{0N} = E_{0N} + q_N(i)/h \, ,$$

$$H^*WH = \begin{bmatrix} W_{NN} & W_{NT} \\ W_{TN} & W_{TT} \end{bmatrix} .$$

Due to the fact that W is symmetric positive definite, when H is an *injection*, H*WH is also symmetric positive definite, as well as W_{TT} and W_{NN}. Therefore, for any tentative value of $h\mathcal{R}_N$, the variational inequality (3.8) has a unique solution $h\mathcal{R}_T$ and, for any tentative value of $h\mathcal{R}_T$, the variational inequality (3.9) has a unique solution $h\mathcal{R}_N$.

The case of the shock law $\mathcal{U}^+_N = -\, e\, \mathcal{U}^-_N$. The following conditions are to be joined to the unilateral conditions (3.5)

If $h\mathcal{R}_N(i-1) \leq 0$ and $h\mathcal{R}_N(i) > 0$, then

(3.5') $h\mathcal{R}_N(i+1) \geq 0$ and $\forall \mathcal{S}_N \geq 0 \;\; (\mathcal{S}_N - h\mathcal{R}_N(i+1)).(q_N(i+1)/h + e\,\mathcal{U}_N(i-1)) \geq 0$,

When, in particular, a shock is expected to occur in the interval (t_i, t_{i+1}), i.e. when

(3.5'') $\qquad h\mathcal{R}_N(i-1) \leq 0$, $h\mathcal{R}_N(i) > 0$, $q_N(i+1) = 0$,

then, taking into account (3.4), one obtains

$$h\mathcal{R}_N(i+1) > 0 \;\Rightarrow\; \mathcal{U}_N(i+1) = -\, e\, \mathcal{U}_N(i-1) \, ,$$

so that the shock law is satisfied within two time steps. The relation (3.5') is the same as (3.5), except that a supplementary term $e\,\mathcal{U}_N(i-1)$ is introduced as soon as (3.5'') is checked to be true. One obtains a system of two coupled variational inequalities similar to (3.8) (3.9).

It should be acknowledged that the above way of accounting for friction in the collision process is rather crude. But it is a fact of life that, in practically all engineering situations, the physical data needed for an accurate analysis of frictional shock are unavailable. So one has to be content with moderately precise prediction of the motion. The numerical procedure described above has the merit of being consistent and stable.

Some remarks on the system (3.8)(3.9). 1° C. Mellouki-Filali [7], has given sufficient, and in some cases necessary, conditions for the existence and uniqueness of the solution to (3.8)(3.9). These conditions involve the friction coefficient and the terms expressing coupling between tangential and normal inertia (terms such as $W_{NT}W_{TT}^{-1}$, $W_{TN}W_{NN}^{-1}$). These conditions are similar to those found in the literature on the existence of solutions to contact problems. They are essentially meant to prevent the occurence of *dynamical locking*, an experimentally observed phenomenon consisting of velocity jumps not induced by collision.When the matrix W is diagonal (such is the case for systems of material points or systems of balls) existence and uniqueness of solutions is always ensured.

2° Various iterative methods may be used to solve (3.8) (3.9), such as Gauss-Seidel with projection, gradient's method, Rosen methods, ...

3° *The case of kinematically dependent constraints.* This shows little difference with the general case except that the matrix H is no more injective. Consequently H*WH is not positive definite but only semi-definite so uniqueness is not ensured. When iterative procedures are applied to solving (3.8) (3.9), the values produced for the local reactions may not converge; however, their global representative $R = H\mathcal{R}$ is found to converge; this at least has been proved in the case of the gradient method. As for $\mathcal{R}$, examples may be exhibited where the succesive approximants converge to a function depending on the data used at the first iteration.

4. AN EXAMPLE.

Ten balls are thrown into a corner and submitted to gravity. Calculation is actually two-dimensional, so balls could also be viewed as disks. The coefficient of restitution equals 0.75 and the friction coefficient equals 0.1. An equilibrium state is eventually reached, two examples of which are shown in Figures 4 and 5. The final rest configuration is very sensitive to small changes in initial conditions. A demonstration software for Macintosh microcomputers may be obtained on request.

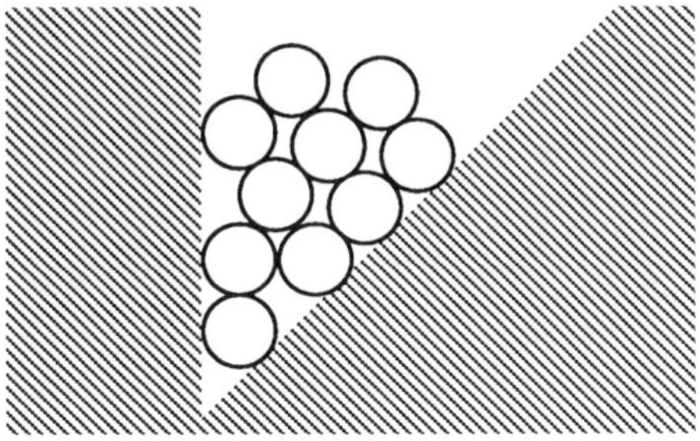

Fig. 4

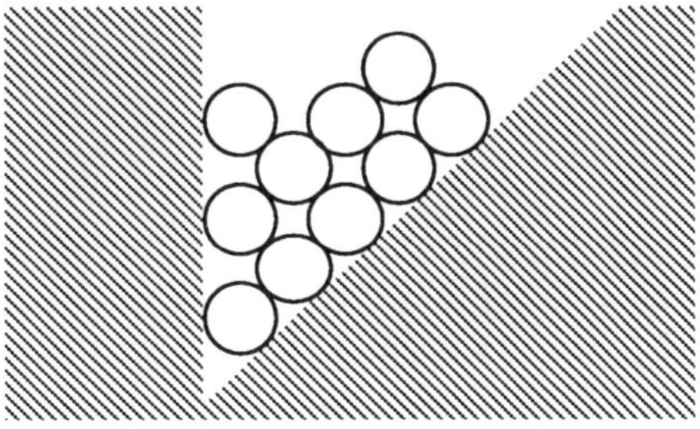

Fig. 5

REFERENCES

[1] MOREAU, J. J.: *Standard inelastic shocks and the dynamics of unilateral constraints*, in: (Ed. G. Del Piero and F. Maceri) *UNILATERAL PROBLEMS IN STRUCTURAL ANALYSIS*, CISM Courses and Lectures N° **288**, Springer-Verlag, Wien, New York 1985.

[2] JEAN, M.: *Unilateral contact with dry friction: time and space discrete variable formulation* , Archiwum Mechanik Stosowanej, Actes du 6ème Symposium Franco Polonais sur la mécanique non linéaire, Villard-de-Lans, 28 Septembre-1er Octobre 1987.

[3] JEAN, M.: *Vibrations d'une tige élastique dans un guide circulaire* , Communication au 9ème Congrès Français de Mécanique, Metz, 5-8 Septembre 1989.

[4] MOREAU, J. J.: *Unilateral contact and dry friction in finite freedom dynamics*, in: (Ed. J. J. Moreau and P. D. Panagiotopoulos) *NONSMOOTH MECHANICS AND APPLICATIONS* , CISM Courses and Lectures, N° **302**, Springer-Verlag, Wien, New York 1988.

[5] JEAN, M. and MOREAU, J. J.: *Dynamics in the presence of unilateral contacts and dry friction; a numerical approach*, in: (Ed. G. Del Piero and F. Maceri) *UNILATERAL PROBLEMS IN STRUCTURAL ANALYSIS - 2* , CISM Courses and Lectures, N° **304**, Springer-Verlag, Wien, New York 1987.

[6] BRACH, R. M.: *Rigid body Collisions* , Journal of Applied Mechanics , 56, 1989, p.133.

[7] MELLOUKI-FILALI, C.: *Problèmes des milieux continus en contact avec frottement : stabilité et convergence des algorithmes numériques*, Thèse de 3ème cycle, Montpellier, Juin 1988.

[8] JOHNSON, K. L.: *CONTACT MECHANICS* , Cambridge Univ. Press, 1985.

[9] MOREAU, J. J.: An expression of classical dynamics, in: (Ed. H. Attouch, J.-P. Aubin, F. Clarke, I. Ekeland) *ANALYSE NON LINEAIRE* , Gauthier-Villars, Paris, 1989, p.1-48.

Laboratoire de Mécanique Générale des Milieux Continus,
CC 051, Université Montpellier 2,
34095 Montpellier Cedex 5, France.

International Series of Numerical Mathematics, Vol. 101, © 1991 Birkhäuser Verlag Basel

CALCULATION OF THE EVOLUTION OF THE FORM OF A RAILWAY WHEEL PROFILE THROUGH WEAR

by J.J. Kalker and A. Chudzikiewicz

SUMMARY

In this paper a method of calculation is presented for the determination of the evolution by wear of railway wheels. The results are compared with the development of the profile of the wheels in the Amsterdam Metro, and a satisfactory agreement is achieved. In ordinary railway lines the agreement is essentially worse, since we did not take into account the wear due to flanging.

INTRODUCTION

Wear determines the life span of a construction. There are several types of wear; we mention

1. Change of form by macroplasticity.
2. Crack formation, as it occurs in fatigue and abrasion. Fatigue is the effect of failure after often repeated loading. Abrasion is the effect that microscopically small particles are pried loose from the contacting surfaces, and are transported out of the contact area.

Wear has been the object of study ever since mechanical constructions have been built. Two famous texts are

[1] E. Rabinowicz, Friction and Wear of Materials, Wiley NY. (1965), and

[2] I.V. Kragelskii, Friction and Wear, Battersworth, London (1965).

Archard [3] pointed out the connection between friction and microplasticity, that is the plasticity on the level of the so–called asperities, *i.e.* the roughness that every metallic surface possesses, and which are of the order of 1 μm in height. After repeated loading, particles are pried loose from the asperity tips, so that it will cause no surprise that *the abrasion wear is taken to be proportional to the frictional work.*

The question arises how the form of a surface changes when it is loaded repeatedly by contact loadings. To answer that question we need a contact progam which, on the basis of the form of the contacting surfaces and their relative motion, calculates the frictional work in the contact area. When that is done, the form of the surfaces is updated by means of the proportionality law of wear and friction. As a consequence of the change in surface form not only the wear changes, but also the motion characteristics of the contacting bodies may be modified.

These effects are present in a pronounced way in the wear phenomena of railway wheels. Several types of wear are recognizable; first, unroundness may arise, in which the wheel is no more exactly a body of revolution but may exhibit waves in the circumferential direction. This is a particularly bothersome form of wear, as the rolling stock is severely and dynamically loaded. Unfortunately the mechanisms that cause unroundness are still unknown, so that it is difficult to perform calculations regarding it. Secondly, we have the form of wear in which the wheel remains a body of revolution, but that in the meridional direction it changes form; we say that the *profile* changes. It is this form of wear that we will consider.

THE CHOICE OF THE ROLLING CONTACT CODE.

In order to calculate the wear, we must have a deep insight in the frictional work. This insight is gained from a *rolling contact code.* The choice of the contact program is important because the various contact codes possess a great variety of properties. Many rolling contact codes have been developed in the course of the years; the most successful ones are

1. The linear theory [4];
2. The theory of Shen, Hedrick and Elkins [5];
3. The complete theory, implementation CONTACT [6];
4. The simplified theory, implementation FASTSIM [7].

The LINEAR THEORY [4] is based on three principles:

A. It is based on Hooke's Law;
B. There is no slip in the contact zone;
C. Its results are accurately known only for elliptical contact zones; they are given in tabulated form in [4].

As a consequence of B. it cannot be used in wear calculations, since the slip enters into the frictional work together with the tangential traction; and the linear theory assumes the slip out of existence.

The THEORY OF SHEN, HEDRICK AND ELKINS [5] is, essentially, a curve fitting theory for the total tangential force in its dependence of the motion characteristics of the vehicle. It was validated with the aid of the linear theory [4] and

CONTACT [6]. It does not yield the local slip nor the local traction, and is therefore not suitable for wear calculations.

The COMPLETE THEORY, as it is implemented in CONTACT [6], is based on three princples.

a. It is based on Hooke's Law.

b. It is based on the friction law of Coulomb. This law is attacked by tribologists, but no satisfactory alternative has as yet been found. In any case there can be a slip velocity under Coulomb's Law, and therefore a frictional work.

c. It is very slow, say 1000× slower than the simplified theory, as it is implemented by FASTSIM [7].

It will be understood that the low speed of CONTACT impairs the accuracy that can be achieved. We hope to accelerate the program in the future by a factor of, say, 50, but then FASTSIM is still 20× as fast.

The SIMPLIFIED THEORY, implemented by FASTSIM [7] is also based on three principles. They are:

i. The *displacement* is taken proportional to the tangential traction at the surface. This differs from Hooke, which takes the *strain* everywhere proportional to the stress.

ii. It is based on Coulomb's Law, see b.

iii. It is very fast, see c.

The approximative character of the assumption (i) impairs the accuracy that can be achieved. Moreover, the linear theory is used to find the material constant, *viz.* the constant of proportionality between displacement and surface load, see (i). It turns out that then the total force is given with a 10–15 % accuracy.

It is also necessary to validate the simplified theory as applied to wear problems. To that end, a comparison was made with CONTACT. This was done in the following way. We calculated a certain wear problem with the aid of two wear programs, which differed only in that one calculated the frictional work with the aid of CONTACT, and the other with the aid of FASTSIM. We found deviations of about 10 % in the form of the worn surfaces, and of about 25 % in the constant of proportionality between wear and frictional work. In view of the crudeness of the wear hypothesis this seems an acceptable result.

We conclude that FASTSIM can be used very well for wear calculations, and therefore we availed ourselves of it.

As we remarked before, the simplified theory utilises the results of the linear theory. These results are known with great accuracy only for elliptical contact areas. Therefore the simplified theory can be applied in practice when the contact area is elliptic. In railway practice, the contact area is not elliptic; it should be *ellipticised.* We will later enter into this aspect of the problem. At this moment we show a figure made by Le The [8] (Fig. 1.). This figure shows, to the left, a wheel set on a track, and on the right side the contact areas are shown. The lower half of each picture shows the real contact area, and the upper half shows the elliptisation (broken line).

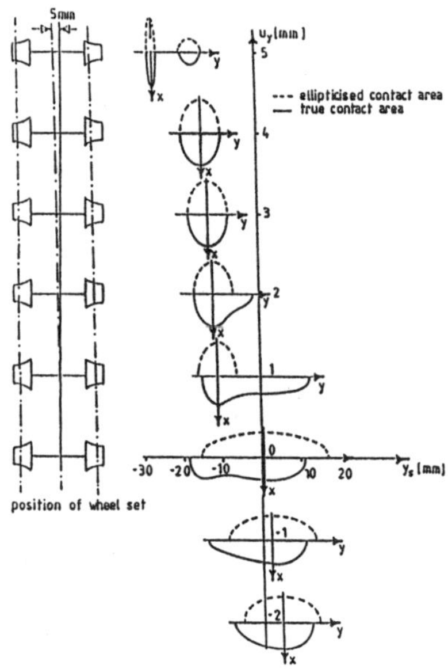

Fig. 1. Contact point, elliptical and non–elliptical contact zones at various lateral positions of the wheel (Type: ORE 1002, UIC 60, Load 60 000 N)
According to Le The [8].

POSITION, FORM AND SIZE OF THE CONTACT AREA

As we see from Fig. 1, the position and form of the contact zone depend upon the position of the contact point on the track. The contact point is the center of the elliptical contact area. It moves to and fro over the rail head and over the wheel profile. Its position depends in a complicated manner on the position of the wheel set on the track, *viz.* on the lateral coordinate q and the yaw angle α , see Fig. 2. We illustrate the position of the contact point K in Fig. 3.

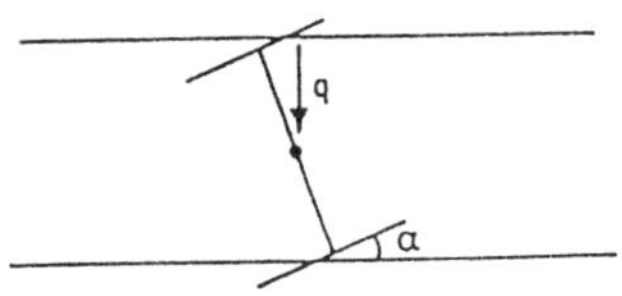

Fig. 2. Wheel set on track.

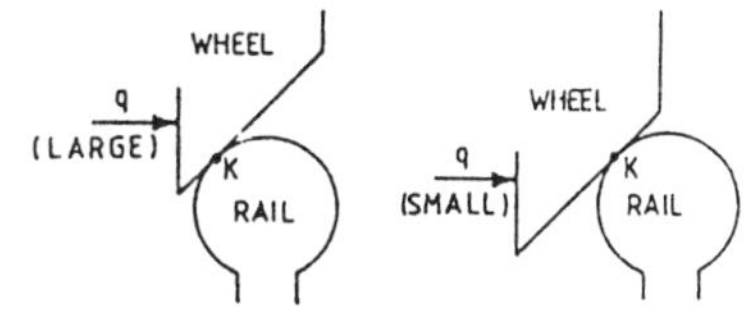

Fig. 3. The contact point K.

The form of the elliptic contact area depends on the form of wheel and rail, and on the normal force. In particular, the curvatures of the wheel and rail surfaces in the contact point determine the form of the elliptic contact area. The contact point itself determines the position; the normal force N determines the size, all according to the Hertz theory [4].

The curvatures and the contact point are determined numerically from the profiles with the aid of a *least squares method.* In this method, wheel and rail are approximated by surfaces of the second degree (paraboloids) in the neighborhood of the contact point, in the following way:

> The profiles of wheel and rail are given at discrete points. One finds an approximation of the contact point by a discrete minimisation of the distance between these points; then one performs a least squares approximation of the profiles based on the points of the profiles near the approximated contact point; from that one finds the curvatures, and an improved estimate of the contact point, by minimisation of the distance between the profiles, which now have a continuous representation.

THE CONTROL OF THE CONTACT PROGRAMS

FASTSIM is controlled by the contact area, the normal compressive force, and the so–called *creepages.* The contact area is elliptic. Contact area and normal pressure are according to the Hertz theory. The normal pressure enters the Coulomb law as a bound on the tangential traction. The creepages are parameters that determine the relative motion of the wheel with respect to the rail, see Fig. 4.

The difference between rolling velocity of the vehicle and the circumferential velocity of the wheel determines the *longitudinal creepage* ξ . This quantity is also influenced by the yaw angular velocity $\dot{\alpha}$.

The lateral velocity $\dot{q}$ and the yaw angle α determine the *lateral creepage* η . The *spin creepage* φ is determined by the angle that the normal on wheel and rail at the contact point makes with the vertical, and by the yaw angle velocity $\dot{\alpha}$.

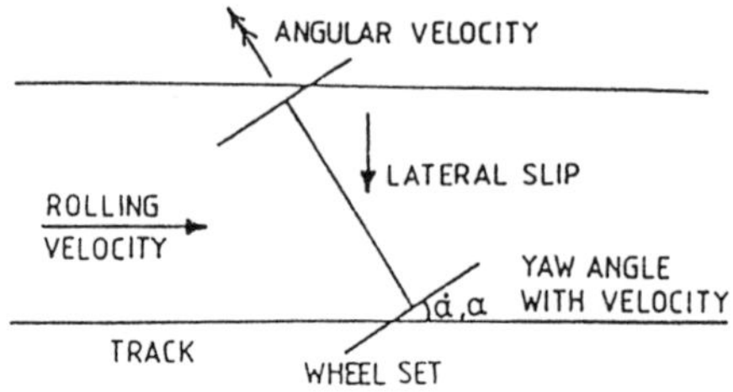

Fig. 4. The state of motion of the wheel set.

THE FRICTIONAL WORK

Consider a particle T .At the time t′ the contact area has the center K′ (Contact point). At the time t the contact area has the center K , see Fig. 5. During the traversal of so small a distance the contact area hardly changes. So one can keep the contact area without alteration, while the particle runs through it, see Fig. 6. From that we see that during its transit of the contact area the particle stores the *sum* of all frictional work from A to B, in the form of wear:

$$\Delta W(T) = \int_A^B \mathbf{p}.\mathbf{v}\, dt = \int_A^B \mathbf{p}.\mathbf{v}\, dx/V,$$

where

- **p**: Tangential traction, a vector
- **v**: Relative velocity, wheel–rail, a vector;
- x: Distance traversed by the wheel over the rail;
- V: Rolling velocity;
- ΔW(T): Frictional work per unit surface which in one cycle (= one revolution of the wheel) is stored in Particle T in the form of wear.

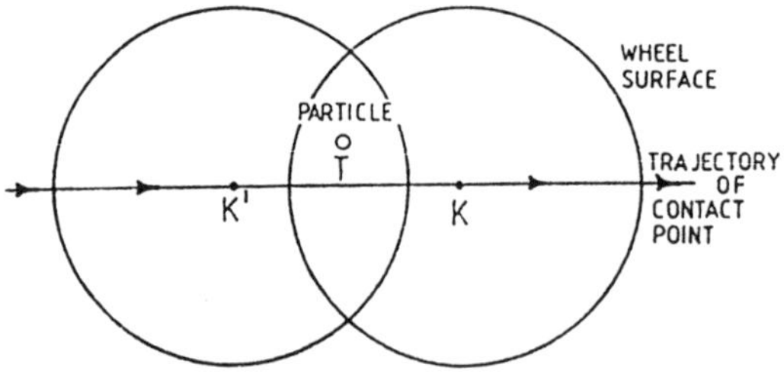

Fig. 5. Trajectory of the contact area over the wheel or the rail.

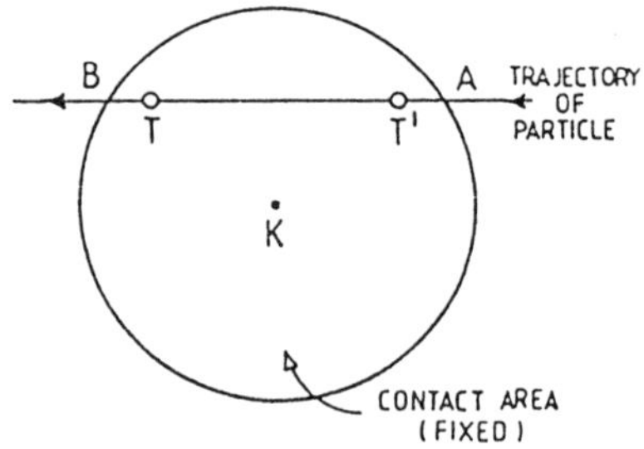

Fig. 6. Trajectory of the particle T through the contact zone.

The particle leaves the contact zone, and meets the rail again *after a full revolution of the wheel.* Then, however, the position, form and size of the contact area can have changed considerably, see Fig. 7, in which only the effect of a change of the position of the contact point is shown. In Position $\mathscr{B}$, once more frictional work is done on the particle T and its surroundings, but the amount differs from that in Position $\mathscr{A}$.

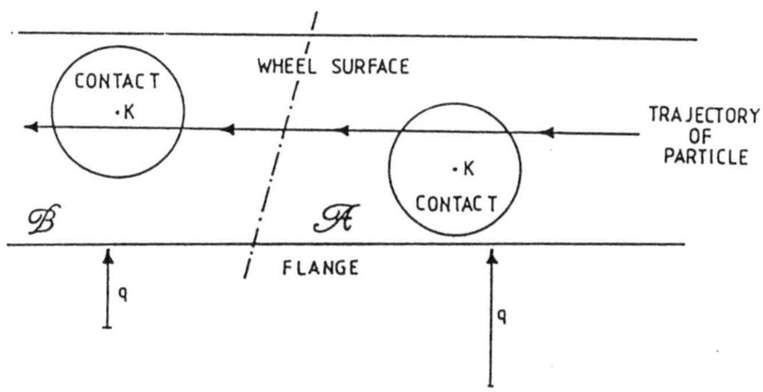

Fig. 7. The position of the contact after n revolutions of the wheel ($\mathscr{A}$) and after m revolutions of the wheel ($\mathscr{B}$).

We consider the increase of the frictional work when a particle traverses a contact zone. It depends on

1. The position y of the particle considered on the the wheel surface with respect to the contact area. We considered that effect above.
2. The size, form and position of the contact. These quantities are uniquely determined by the lateral coordinate q, see Fig. 7, and by the profiles of wheel and rail.
3. The local motion of the wheel relative to the rail, as it is determined by the creepages ξ,η,φ.

The profiles change very slowly as a consequence of wear. Typical is 1 μm per 33 000 Revolutions = per 100 km distance traversed (diameter wheel: 1 m).
We will take together so many revolutions that the profiles change just a little, and consider that a single "unit wear step". So we find:

$\Delta W(y) = f(y,q)$, f unchanged in a single unit wear step.

THE FRICTIONAL WORK IN A UNIT STEP

As we said, the wear is about 1 μm per 33 000 revolutions of the wheel. 33 000 Rev = 100 000 m = 100 km rolling appears to be too large a step to serve as a "unit wear step". Such a step must satisfy two criteria:

- If it is too large, the solution may become inaccurate, and unstable;
- If it is too small, the processing time is unnecessarily large.

In our work, we found that a unit step of

M = 13300 REV = 40 km traversed in rolling

gave just the good accuracy.

After we have analysed a unit step, we traverse, say,

7500 steps=7500 × 40 km =300 000 km.

In order to calculate the wear in a unit step, we can simulate the motion in a unit step, either by calculation or by measurement. Then we can analyse each revolution separately, and sum over all 13300 contributions. It is also possible to calculate the *probability* $P(q_i,\Delta q)$ *that* q *lies between* q_i *and* $q_i+\Delta q$. Then the expectation of the frictional work is, per revolution

$$E(\Delta W(y)) = \sum_i P(q_i,\Delta q)\, f(y,q_i), \quad q_i = i\Delta q$$

and the total frictional work in a unit step is, with sufficient precision

$$\Delta W_{U.S.}(y) = 13300\ E(\Delta W(y))$$

THE CALCULATION OF THE PROBABILITY $P(q_i,\Delta q)$.

As we saw before, $\Delta W(y)$ follows from contact mechanical considerations. We now determine the probability distribution $P(q_i,\Delta q)$. To that end we consider the lateral movement of the wheel over a large distance X, e.g. the distance traversed during a unit wear step, see Fig. 8.

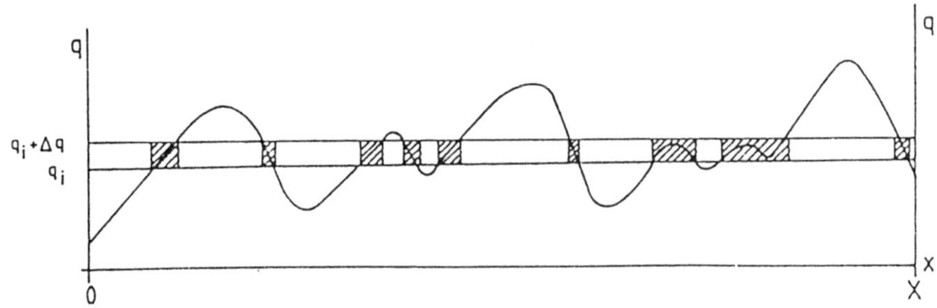

Fig. 8. The calculation of the probability $P(q_i,\Delta q)$.

Let the total length of the intervals where $q_i \leq q \leq q_i+\Delta q$ be x_i; then

$$P(q_i,\Delta q) \approx x_i/X.$$

So the frictional work in a unit step becomes

$$\Delta W_{U.S.}(y) = 13300\ E(\Delta W(y)),$$

$$E(\Delta W(y)) = \sum_i \frac{x_i}{X} f(y,q_i)\ ; \quad q_i = i\Delta q.$$

In practice it appears that about 20 values for q suffice to obtain satisfactory results. From this it appears that this probabilistic method is much more efficient than a method of direct simulation.

There are two different ways to find the trajectory of the wheel over the rail:

1. SIMULATION – that is, calculation of the motion of the vehicle, with the following inputs: vehicle characteristics, track characteristics and irregularities, braking and accelerating forces, and the running velocity.

Such programs have been made by many Institutes; however, a good simulation program is extremely complicated, and therefore hard to validate.

2. MEASUREMENT – With the aid of a special measuring vehicle the motion may be measured directly.

Netherlands Railways possesses such a vehicle which can store the motion over 100 km. It is the intention to connect this measurement to our wear program, but this has not yet been done.

It is seen that the investigation is not closed yet, as neither of these possibilities has been realised to my knowledge. What we have done is a pilot investigation in which the calculation of the probabilities $P(q_i,\Delta q)$ is drastically simplified. Indeed, we consider a stochastic motion of the following type:

$$q = v' + v + a \sin(\omega x),$$

where

- q is the lateral displacement;
- v' is a constant;
- v is a stochastic variable, uniformly distributed between v_1 and v_2, with $0<v_1<v_2$;
- a is a stochastic variable uniformly distributed between 0 and v;
- ω is a fixed angular frequency, dimenstion m^{-1};
- x is the coordinate along the track.

The probability $P(q_i,\Delta q)$ can now be calculated. Owing to the uniformity of the distributions most of these calculations can be performed by elementary means; only one numerical integration over a single variable is involved [9]. To have an idea of the magnitude of the quantities involved, we give some numerical values:

$v' = 0$ mm;

$v_1 = 2$ mm;

$v_2 = 8$ mm;

ω = immaterial (mm^{-1});

N Total compressive force, 60 000 N per contact;

R wheel Radius, 500 mm;

Contact area, typically a circle with radius 6 mm, see fig. 1;

E Young's modulus, $2.1_{10}5$ N/mm^2;

ν Poisson's Ratio, 0.28 (dimensionless);

ξ Longuitudinal creepage, of the order of 0.0003 (dimensionless);

η Lateral creepage, of the order of 0.0005 (dimensionless);

φ Spin creepage, of the order of $2.5_{10}{-5}$ mm^{-1}.

THE RATE OF WEAR

Wear can be defined in two ways, which are equivalent. The first way is the more general definition. It reads

a. The mass of material worn away – dimension: kg, or, more appropriately, $\mu g = {}_{10}{-9}$ kg.

The second definition is more suitable for the calculation of the form of the worn surface. It reads:

b. The depth of the part that was worn away (dimension: mm).

In a., the wear is proportional to the frictional work performed on the particle. The dimension of the frictional work is: N mm . In b., the wear is proportional to the frictional work that has been performed on the unit of surface. Its dimension is N/mm.

We are interested in the constants of proportionality:

Mass worn away (μg) = C×Frictional work (N mm)

⇒ Dimension of C: μg/(N mm)

Depth worn away (mm) = C′×Frictional work/unit area (N/mm)

⇒ Dimension of C′: mm^2/N = mm^3/(N mm).

The constants C and C′ convert in the following manner:

$$C' \text{ (in mm}^2\text{/N)} = \frac{1}{\rho} C,$$

C in μg/(N mm), ρ: density = 8000 μg/mm^3 = 8000 kg/m^3.

In our work, we found that

$C \approx 0.00124\ \mu g/(N\ mm)$;

$C' \approx 1.55_{10}{-7}\ mm^2/N$.

The rate of wear was measured by Mr. H. Lehna of the RWTH Aachen, Germany [10]. He found, after very careful measurements:

$1_{10}{-4}\ \mu g/(N\ mm) < C < 1_{10}{-2}\ \mu g/(N\ mm)$.

In our computer program we analysed one unit wear step in the simplified probabilistic manner of the preceding Section. One unit wear step corresponds to M revolutions of the wheel, or to 3M m traversed, as the diameter of the wheel is roughly 1 m. The constants of proportionality become

$CM = 16.5\ \mu g/(N\ mm)$,

$C'M = 0.0020\ mm^2/N$.

These (equivalent) values are determined by the behaviour of the program, as one can choose M arbitrarily. CM must be neither too large nor too small:

CM too large results in inaccurate results;

CM too small results in long processing times.

RESULTS

We compare with railway practice. We find from our computer results that

1 mm of wear corresponds to 2500 unit wear steps with the CM of the previous section.

We have learned from the experience of Netherlands Railways that

1 mm wear corresponds to 100 000 km traversed.

So we have:

2500 unit wear steps = 2500 M revolutions of the wheel =2500 M ×3 m traversed = 7.5 M km traversed

= 100 000 km traversed

$\Rightarrow M = 13300,\ C = \{16.5/M\}\ \mu g/(N\ mm) = 1.24\ _{10}{-3}\ \mu g/(N\ mm)$.

It is satisfying to observe that the value of C lies well within the range of Lehna's experiments:

$1_{10}{-4}\ \mu g/(N\ mm) < 1.24\ _{10}{-3}\ \mu g/(N\ mm) < 1_{10}{-2}\ \mu g/(N\ mm)$.

We simulated the wear of a realistic wheel profile on a realistic rail. We assumed that the rail did not wear, but that the wheel did. The results are shown in Fig. 9. In Fig. 9a we show the wheel and the rail, and the form of the wheel after 300 000 km of running. As in each wear step about 20 contact problems are calculated, the total number of contact problems computed is about 150 000. This large number explains, why such importance is attached to the high speed of the rolling contact code. In Fig. 9c we show the evolution of the wear, after 40 000, 100 000, 200 000, and 300 000 km. The vertical scale has been extended by a factor 10 with respect to the horizontal scale, while in Fig. 9a the vertical and the horizontal scales are the same. Fig. 9b is identical to Fig. 9c, but in addition the relative positions of wheel and rail are shown, as well as the position of the contact point at the extreme edges of the range. It is seen that the position of the contact point is greatly influenced by the wear. It is also seen that the form of the worn part of the wheel more or less fits the form of the unchanging rail, irrespective, apparently, of the initial form of the wheel.

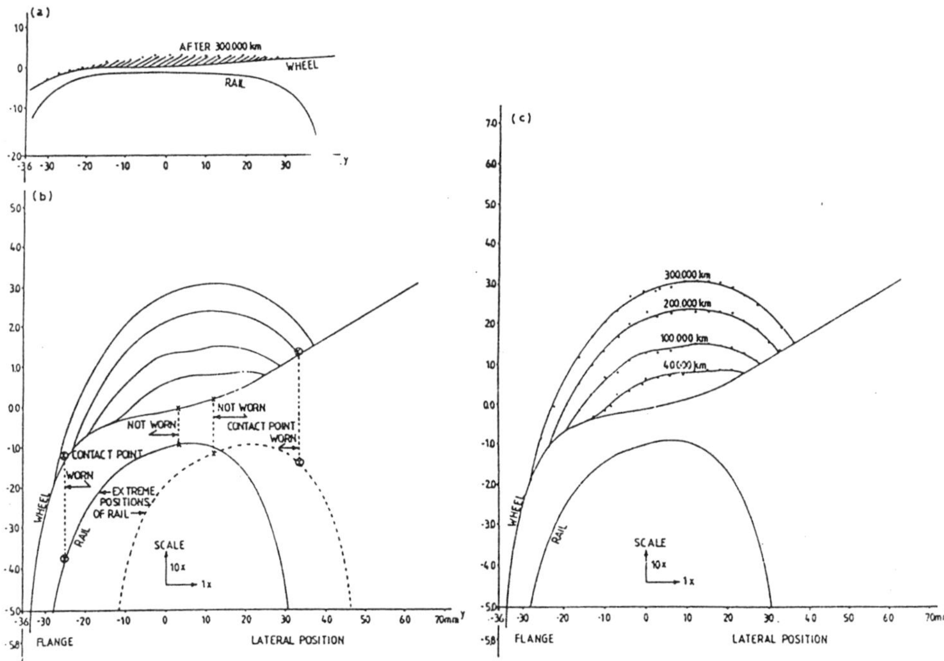

Fig. 9. The results of a simulation of the wear of a railway wheel.

a: Identical scales, horizontal and vertical.

b: Different scales, horizontal and vertical. The extreme relative positions and the contact points are shown.

c: Different scales, horizontal and vertical. Calculated points are shown.

Fig. 9 may be compared with the real worn forms of railway wheels, shown in Fig. 10. Fig. 10a shows the worn wheel of a freight car. It differs considerably from Fig. 9, mainly because of the circumstance that in the Fig. 10a the flange is worn. We did not simulate flange wear in Fig. 9. The imprint of the rail is very clear also in Fig. 10a. In Fig. 10b the worn wheel of a vehicle of the Amsterdam Metro is shown. Wear extends over a much narrower band than in Fig. 10a, and Fig. 10b shows a great similarity to Fig. 9. The reason of the narrower wear band should be sought in the better and more uniform quality of the track of the Amsterdam metro, and in the circumstance that the banking of the curves in the metro track corresponds more closely to the vehicle velocity than under the conditions of Fig. 10a.

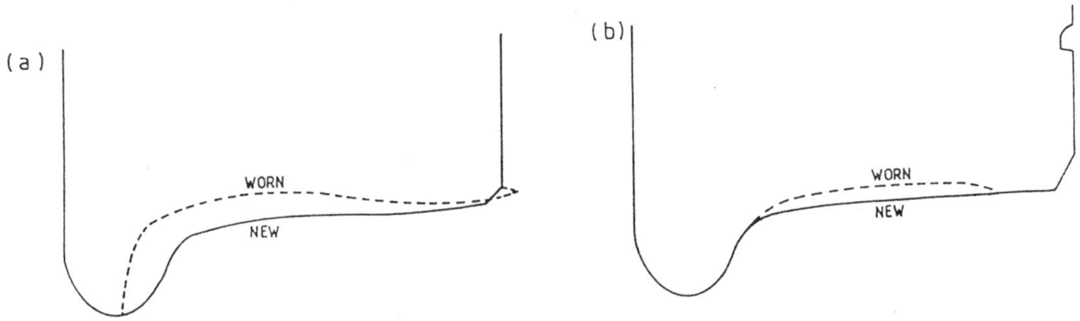

Fig. 10. Worn wheels in practice.
a: A freight car wheel.
b: An Amsterdam Metro wheel.

CONCLUSION

It is shown in this paper how one can perform a wear calculation quickly and yet with sufficient precision. Heart of the calculation is the FASTSIM algorithm that implements the simplified rolling contact theory. We have given preliminary calculations which yield results that are similar to railway practice. It remains to introduce the results of realistic motion simulations or measurements into the program.

REFERENCES

[1] E. Rabinowicz, Friction and Wear of Materials, Wiley, New York (1965).

[2] I.V. Kragelskii, Friction and Wear, Battersworth, London (1965).

[3] J.F. Archard, Elastic deformation and the laws of friction, Proc. Roy. Soc. London Ser. A, **243**, p. 190–205.

[4] J.J. Kalker, On the rolling contact of two elastic bodies in the presence of dry friction, Thesis Delft (1967).

[5] Z.Y. Shen, J.K. Hedrick, J.A. Elkins, A comparison of alternative creep force models for rail vehicle dynamic analysis. In: Tyhe Dynamics of Vehicles, Ed. J.K. Hedrick, Proc. 8th IAVSD Symposium, MIT, Cambridge, MA. Swets and Zeitlinger, Lisse, the Netherlands.

[6] J.J. Kalker, On the contact problem in elastostatics. In: Unilateral problems in structural analysis, Ed. G Del Piero & F. Maceri, CISM courses and lectures 288, Springer Wien–New York (1985)

[7] J.J. Kalker, A fast algorithm for the simplified theory of rolling contact, Vehicle System Dynamics 11 (1982) p. 1–13.

[8] Le The Hung, Normal– und Tangentialspannungsberechnung beim rollenden Kontakt für Rotationskörper mit nichtelliptischen Kontaktflächen. Fortschrittberichte VDI, Reihe 12, Nr. 87 (1987)

[9] J.J. Kalker, Wheel–rail wear calculations with the program CONTACT, in: Contact mechanics and wear of rail/wheel systems II, Eds. Gladwell, Ghonem and Kalousek, p.3–26 (1987).

[10] H. Lehna, Letter to J.J. Kalker, dated 7th March 1989.

Faculty of Engineering Mathematics and Informatics
Delft University of Technology
Julianalaan 132
Delft – The Netherlands

A Global Existence Result for the Quasistatic Frictional Contact Problem with Normal Compliance

by

A. Klarbring
Department of Mechanical Engineering
Linköping Institute of Technology
Linköping, Sweden

A. Mikelić
Department of Theoretical Physics
"Ruder Bošković" Institute
P.O. Box 1016
41001 Zagreb, Yugoslavia

M. Shillor
Department of Mathematical Sciences
Oakland University
Rochester, MI 48309, USA

ABSTRACT

We consider the quasistatic problem of the contact of an elastic body with a rigid foundation in the presence of friction. The contact condition is taken as a power law normal compliance. We prove, for forces and initial data that are not too large, the existence of a solution $\mathbf{u}$ such that $\mathbf{u} \in C([0,T];\mathbf{H}^1(\Omega))$ and $d\mathbf{u}/dt \in L^2(0,T;\mathbf{H}^1(\Omega))$. The main tools are from the theory of differential inclusions.

Key words: nonlinear parabolic variational inequalities, quasistatic frictional problems, surface normal compliance.

§*1. Introduction*

We consider the problem of quasistatic contact of a linearly elastic body with a rigid foundation in the presence of dry friction and surface normal compliance. We prove the existence of a global solution when the loads or the contact coefficients are not too large and when the friction law is regularized in time. For the linearized problem, of interest on its own, we prove the existence of a unique solution without any regularization or restrictions on the size of the loads or contact coefficients .

We use the power law surface normal compliance for the description of the contact of the body with the foundation. This law was recently adopted by Oden and Martins [OM] based on an impressive analysis of experimental results and engineering literature. It was studied in the papers by [MO], [RMOC], [RO], [EMS] and [KMS1, 2, 3, 4] and can be found in Kikuchi and Oden [KO]. Most of these papers deal with the static problems except for [OM] and [MO] where the dynamic cases with viscous damping were considered. The main motivation for the introduction of the viscous damping was as a regularization of the problem in time.

The quasistatic problem of the frictional contact of an elastic body with a rigid foundation and a power law surface normal compliance was considered by Klarbring, Mikelić and Shillor [KMS1]. We derived two opproximations of the full problem, the incremental and the rate problems, but the existence of a solution to the quasistatic problem itself was left open. Here we establish the existence of a strong solution without the need of the viscous regularization.

We follow [KMS1] in the formulation of the mathematical model. A similar approach can be found in Oden and Martins [OM] and in Kikuchi and Oden [KO, Ch11].

Suppose that the elastic body occupies the region $\Omega \subset R^n$ ($n = 2$ or 3) and can come into contact with a rigid foundation over a part Γ_C of its boundary $\partial\Omega$. Let $\partial\Omega = \overline{\Gamma}_U \cup \overline{\Gamma}_C$ and assume that Γ_U and Γ_C are smooth (say $C^{1,1}$) $n-1$ dimensional manifolds. The body is subjected to body forces of (volume) density $\mathbf{f}$ and is held fixed over Γ_U. Let

$\mathbf{u} = \mathbf{u}(\mathbf{x},t) = (u_1(\mathbf{x},t),\ldots,u_n(\mathbf{x},t))$ denote the displacement of the point $\mathbf{x} \in \Omega$ at time $t \in [0,T]$ and let $\sigma = \sigma(\mathbf{x},t)$ be the stress tensor.

The quasistatic problem is to find $\{\mathbf{u},\sigma\}$, $\mathbf{u} : \Omega \times [0,T] \to R^n$, $\sigma : \Omega \times [0,T] \to R^{n^2}$, $\sigma = (\sigma_{ij})$, $\sigma_{ij} = \sigma_{ji}$, such that

$$\text{div } \sigma + \mathbf{f} = 0, \qquad \text{in } \Omega, \quad 0 \leq t \leq T, \tag{1.1}$$

$$\sigma_{ij} = \sigma_{ij}(\mathbf{u}) = a_{ijk\ell}\frac{\partial u_k}{\partial x_\ell}, \qquad \text{in } \Omega, \quad 0 \leq t \leq T, \tag{1.2}$$

$$\mathbf{u} = 0, \qquad \text{on } \Gamma_U, \quad 0 \leq t \leq T, \tag{1.3}$$

$$\mathbf{u} = \mathbf{u}_0, \qquad \text{in } \Omega, \quad t = 0\ . \tag{1.4}$$

These are the equations of linear elasticity with Dirichlet boundary condition on Γ_U. The contact conditions, which represent the interaction between the body and the foundation over the contact surface Γ_C are

$$-\sigma_N = C_N(u_N - g_*)_+^{m_N}, \qquad \text{on } \Gamma_C, \quad 0 \leq t \leq T, \tag{1.5}$$

$$|\sigma_T| \leq C_T(u_N - g_*)_+^{m_T}, \qquad \text{on } \Gamma_C, \quad 0 \leq t \leq T, \tag{1.6}$$

and

$$|\sigma_T| < C_T(u_N - g_*)_+^{m_T} \ \Rightarrow\ d\mathbf{u}_T/dt = 0,$$

$$|\sigma_T| = C_T(u_N - g_*)_+^{m_T} \ \Rightarrow\ d\mathbf{u}_T/dt = -\lambda\sigma_T, \text{ some } \lambda \geq 0.$$

Here (1.5) is the power law *surface normal compliance* and (1.6) is the modified Coulomb's law of dry friction. Notice that Coulomb's law is related to the rate of tangential displacement and not to the displacement itself i.e. to $d\mathbf{u}_T/dt$ and not to $\mathbf{u}_T$. The following notation is used above: if $\mathbf{n} = (n_1,\ldots,n_n)$ is the outward unit normal vector on Γ_C then

$$\sigma_N = \sigma_{ij}n_i n_j, \qquad \sigma_T = \sigma\mathbf{n} - \sigma_N\mathbf{n},$$

$$u_N = u_i n_i, \qquad \mathbf{u}_T = \mathbf{u} - u_N\mathbf{n}$$

and summation is implied over repeated indices, $i,j = 1,\ldots,n$.

Note that, on the one hand (1.5) may be thought of as representing a substitution of the rigid foundation by nonlinear springs. But, on the other hand, since machined metallic surfaces posses undulations, the so called asperities, (1.5) may represent the elastic behaviour of these asperities and the formulation may still be thought of as a rigid foundation. The latter interpretation assings some physical properties to the contact surfact Γ_C and the g_* represents the initial gap between the tips of the asperities and the rigid foundation. We assume $g_* \in L^\infty(\Gamma_C)$. As usual the system is taken as elliptic i.e.

$$a_{ijk\ell}\, \xi_{ij}\, \xi_{k\ell} \;\geq \alpha|\xi|^2 \tag{1.7}$$

for each symmetric matrix ξ $n \times n$ (i.e. $\xi_{ij} = \xi_{ji}$) for some $\alpha > 0$ and we assume that

$$a_{ijk\ell} = a_{jik\ell} = a_{k\ell ij} \in C^1(\Omega).$$

Next consider C_T and C_N which are nonnegative functions on Γ_C and represent the physics of friction. It is assumed that $C_T, C_N \in L^\infty(\Gamma_C)$. The exponents m_N and m_T are experimental coefficients.

The problem (1.1) - (1.6) was considered in [KMS1] where it was discretized in time and we proved the existence of a solution in each time step. In order to consider the existence of a solution to the quasistatic problem we introduce a modification of (1.5) and (1.6). From the mathematical point of view this is a regularization in time of the contact stresses, necessary for our *a* priori estimates. Physically it may represent a delay related to the processes on the contact surface. So we rewrite (1.5) and (1.6) as follows

$$\begin{aligned} &-\sigma_N = g_N, \\ &|\sigma_T| \leq g_T, \\ |\sigma_T| < g_T \Rightarrow \qquad &\qquad \dot{\mathbf{u}}_T = 0, \\ |\sigma_T| = g_T \Rightarrow \qquad &\qquad \dot{\mathbf{u}}_T = -\lambda\sigma_T, \qquad \text{some } \lambda \geq 0. \end{aligned} \tag{1.8}$$

where $\dot{\mathbf{u}} = d\mathbf{u}/dt$. These hold on Γ_C for $0 \leq t \leq T$. Our regularization is in the modification of g_N and g_T that we proceed to describe. But first we remark that the dynamical problem with Coulomb's friction with $g_N = g_N(\mathbf{x})$ and $g_T = g_T(\mathbf{x})$ was considered in Duvaut and Lions [DL, pp158-162] where they remark that they are unable to solve the problem with time dependent g_N and g_T by using a combination of regularization of the nondifferentiable functional and Galerkin's method. We on the other hand consider the case $g_N = g_N(\mathbf{x}, t)$ and $g_T = g_T(\mathbf{x}, t)$ in Section 2 and prove the existence of a unique solution using the theory of [ABDP], [AD1], [Pel] and [Pe2].

We turn to describe the regularization of (1.5) and (1.6). The physical motivation for introducing time smoothing is as follows, and resembles to some extent the reasons that led Oden and Pires [OP] (see also [KO]) to introduce their nonlocal law of friction. In (1.5) and (1.6) g_N and g_T are the contact pressure and the friction bound, respectively, and both depend instantaneously on the normal displacement. But according to the well-known friction theory of Bowden and Tabor [BT] frictional resistence is caused by the shearing of junctions that are formed by the welding of the surface undulations. The welding cannot be assumed instantaneous neither the shearing and breaking of these welded junctions. Therefore the introduction of a time delay seems to be reasonable from the physical point of view and indeed makes both g_N and g_T depend on the history.

We use the following smoothing. Let $\rho \in C_0^\infty(R)$ and in order to observe causality $\rho(s) = 0$ if $s \leq 0$ and such that $\operatorname{supp}\rho \subset [0, \epsilon]$ for an arbitrary but fixed $\epsilon > 0$. Our method does not depend on a particular form of ρ but to fix ideas we may choose

$$\rho(s) = \rho_\epsilon(s) = \begin{cases} 0 & s \leq 0 \text{ or } \epsilon \leq s, \\ \frac{\beta}{\epsilon} \exp\left[-\frac{\epsilon^2}{s(\epsilon - s)}\right] & 0 \leq s \leq \epsilon, \end{cases} \tag{1.9}$$

where

$$\beta^{-1} = \int_0^1 \exp\left[-\frac{1}{s(1-s)}\right] ds.$$

The choice of β is for normalization purposes and it is easy to see that $\{\rho_{\epsilon_n}\}$ is a δ-sequence, i.e. a sequence that converges to the Dirac delta in the sense of distributions, for any sequence $\{\epsilon_n\}$ such that $\epsilon_n \to 0$ as $n \to \infty$.

Then we define

$$g_N(\mathbf{x},t,\mathbf{u}) = \int_0^t C_N(\mathbf{x})(u_N(\tau) - g_*)_+^{m_N} \rho(t-\tau)d\tau + \\ C_N(\mathbf{x})(u_N(0) - g_*)_+^{m_N}[1 - \int_0^t \rho(t-\tau)d\tau], \tag{1.10}$$

and

$$g_T(\mathbf{x},t,\mathbf{u}) = \int_0^t C_T(\mathbf{x})(u_N(\tau) - g_*)_+^{m_T} \rho(t-\tau)d\tau + \\ C_T(\mathbf{x})(u_N(\tau) - g_*)_+^{m_T}[1 - \int_0^t \rho(t-\tau)d\tau] \tag{1.11}$$

We notice that g_N and g_T are smooth in time and also $g_N \to C_N(u_N - g_*)_+^{m_N}$ and $g_T \to C_T(u_N - g_*)_*^{m_T}$ for every $(\mathbf{x},t)$ as $\epsilon \to 0$. The second parts of the right hand sides of (1.10) and (1.11) are introduced for the approximation at times $0 < t \leq \epsilon$. From the definition (1.9) of ρ and from (1.10) and (1.11) it follows that $g_N(t)$ and $g_T(t)$ depend on the history of the process during the time interval $t - \epsilon \leq \tau \leq t$. By appropriate choice of ϵ we may make it as short as we wish.

We define the bilinear form

$$a(\mathbf{u},\mathbf{v}) = \int_\Omega a_{ijk\ell} \frac{\partial u_i}{\partial x_j} \frac{\partial v_k}{\partial x_\ell} dx, \tag{1.12}$$

where summation over repeated indices is implied and let

$$V = \{\mathbf{z} \in \mathbf{H}^1(\Omega);\ \ \mathbf{z} = 0 \text{ on } \Gamma_U\}. \tag{1.13}$$

Then the weak or variational formulation of (1.1) - (1.4), (1.8) with (1.10) and (1.11) is (see e. g. [KMS1])

$$(P) \quad \begin{cases} \text{find } \mathbf{u} \in C([0,T];V) \text{ such that } \mathbf{u}(0) = \mathbf{u}_0 \in V,\ \dot{\mathbf{u}} \in L^2(0,T;V) \text{ and} \\ a(\mathbf{u},\mathbf{v}-\dot{\mathbf{u}}) + \int_{\Gamma_C} g_T(\mathbf{x},t,\mathbf{u})(|\mathbf{v}_T| - |\dot{\mathbf{u}}_T|)ds + \int_{\Gamma_C} g_N(\mathbf{x},t,\mathbf{u})(v_N - \dot{u}_N)ds \geq \\ \geq \int_\Omega \mathbf{f}(\mathbf{v}-\dot{\mathbf{u}})dx, \qquad \forall \mathbf{v} \in V, \text{ a.e. on } (0,T), \end{cases} \tag{1.14}$$

where here and below $\dot{\mathbf{u}} = d\mathbf{u}/dt$. The problem (P) is closely connected to evolution inclusions of the type

$$\frac{du}{dt}(t) + \partial\varphi(t, u(t)) \ni f(t), \tag{1.15}$$

where $\partial\varphi(t,\cdot)$ is the subdifferential of a function $\varphi(t,\cdot)$ defined on a Hilbert space H.

Problems of this type when φ does not depend on t are classical and were extensively studied in Brézis [B 1,3] among others. The more general case when φ depends on the time was considered in [B2], in Attouch and Damlamian [AD 1, 2] and Péralba [Pe 1, 2]. We use their results to prove ours. If for $t \in [0,T]$ $\varphi(t,\cdot) : H \to R$ is a proper convex function and $f \in L^2(0,T;H)$ usually the solution u to (1.15) is to be found in $C([0,T];H) \cap W^{1,2}_{\text{loc}}(0,T;H)$ and is called a strong solution.

We prove the existence of a strong solution to problem (P). To this end we consider the linearized problem (P_L) in Section 2 where $g_N = g_N(\mathbf{x},t)$ and $g_T = g_T(\mathbf{x},t)$, i.e. both independent of $\mathbf{u}$. As was mentioned above, it follows from Duvaut and Lions [DL] see also Panagiotopoulos [Pa], that the classical regularization approach cannot be applied to problems with g_N and g_T depending on time. Therefore using evolution inclusions we are able to close a part of the gap and prove an existence result for the quasistatic problem with prescribed contact pressure and friction bound. Then we use Péralba's results [Pe1] to obtain *a* priori estimates for the solutions of (P_L) that are essential to the proof of the existence of a solution of the full problem (P) in Section 3. There we define an appropriate function G, the fixed point of which is a solution of (P). Since we use Schauders fixed point theorem we obtain global existence either when the data $\mathbf{f}, \mathbf{u}_0$ and g_* are small or the C_N and C_T are small.

The existence of a global solution to the quasistatic problem confirms the conclusions based on many numerical experiments that there is no breakdown of the strong solution. A numerical approach, based on the time discretized quasistatic problem, was discussed in [KMS2].

When this work was finished we received the preprint "A quasistatic frictional problem with normal compliance" by L. E. Andersson [A] where the same problem (P) is considered. He proves the existence of a solution, without the time regularization, for small data or coefficients. His method is completely different from ours and is based on the time discretized problems described in [KMS1]. He obtains the necessary estimates for solutions to each time step and upon passing to the limit as the time step shrinks to zero, he obtains a solution to (P).

Acknowledgement, The authors would like to thank Prof. J. J. Moreau for sending the doctoral thesis of J. C. Péralba.

A part of this work was done when A.M. and M.S. visited the Linköping Institute of Technology during August 1988 and the support from the Swedish Board for Technical Development is gratefully acknowledged. Finally M.S. was supported in part by the Oakland University Faculty Research Award.

§2. Existence for the linearized problem and a priori estimates

In this Section we consider a linearized version of (1.14) and using some results from parabolic variational inequalities we prove the existence and uniqueness of the solution. Then we obtain two *a* priori estimates necessary for the next Section.

So we consider the problem

$$(P_L)\qquad \begin{cases} \text{find } \mathbf{w} \in C([0,T];V) \text{ such that } \mathbf{w}(0) = \mathbf{u}_0, d\mathbf{w}/dt \in L^2(0,T;V) \text{ and} \\ a(\mathbf{w}, \mathbf{v} - \dot{\mathbf{w}}) + \int_{\Gamma_C} g(|\mathbf{v}_T| - |\dot{\mathbf{w}}_T|)ds \geq \int_\Omega \mathbf{f}(\mathbf{v} - \dot{\mathbf{w}})dx + \\ \qquad + \int_{\Gamma_C} F_N(v_N - \dot{w}_N)ds, \quad \forall \mathbf{v} \in V, \text{ a.e. on } (0,T). \end{cases} \tag{2.1}$$

Here and below "$\cdot$" denotes d/dt, $\mathbf{u}_0 \in V$ is given, so is F_N and $g = g(\mathbf{x}, t)$ is independent of $\mathbf{v}$.

As was mentioned in the introduction, this problem has interest of its own. The time independent case $g = g(\mathbf{x})$ was considered in Duvaut and Lions [DL Ch.III, §5] where the existence of the unique solution was proved for the fully dynamic problem. Furthermore they gave an explanation why their method does not apply to a time dependent g [DL, pp156-161]. Similar problems were considered recently by Panagiotopoulos [Pa] again for $g = g(\mathbf{x})$ only. However, using parabolic variational inequalities we prove the existence of a unique solution for the problem (P_L) with time dependent $g = g(\mathbf{x}, t)$.

We turn to rewriting (P_L) in a more convenient form. We introduce a new norm on V.

$$\|\mathbf{v}\|_V^2 = a(\mathbf{v}, \mathbf{v}), \qquad \mathbf{v} \in V. \tag{2.2}$$

By Korn's inequality (e.g. [DL, p110]) this is a norm and is equivalent to the $H^1(\Omega)$ norm on V. Moreover since $a(\cdot,\cdot)$ is a symmetric bilinear form on V the norm $\|\cdot\|_V$ is induced by the inner product

$$(\mathbf{u}, \mathbf{v})_V = a(\mathbf{u}, \mathbf{v}), \qquad \mathbf{u}, \mathbf{v} \in V. \tag{2.3}$$

Next consider the linear form

$$\Lambda(\mathbf{v}) = \int_{\Omega} \mathbf{f}\mathbf{v}dx + \int_{\Gamma_C} F_N v_N ds, \qquad \mathbf{v} \in V. \tag{2.4}$$

By our assumptions $\mathbf{f} \in W^{1,\infty}(0,T;\mathbf{L}^2(\Omega))$ and in addition we assume

$$F_N \in W^{1,\infty}(0,T;\mathbf{L}^{\frac{4}{3}}(\Gamma_C)), \tag{2.5}$$

$$g \in W^{1,\infty}(0,T;\mathbf{L}^{\frac{4}{3}}(\Gamma_C)), \qquad g \geq 0 \text{ a.e. on } \Gamma_C \times (0,T). \tag{2.6}$$

It follows that Λ is a continuous linear functional on $V \times (0,T)$ for $n \leq 3$. Then by the Riesz representation theorem there exists $\mathbf{F} \in W^{1,\infty}(0,T;V)$ such that

$$(\mathbf{F},\mathbf{v})_V = \Lambda(\mathbf{v}), \qquad \forall \mathbf{v} \in V, \text{ a.e. on } (0,T). \tag{2.7}$$

Let us introduce the functional $\varphi : [0,T] \times V \to (-\infty,+\infty]$, defined by

$$\varphi(t,\mathbf{v}) = \int_{\Gamma_C} g(\mathbf{x},t)|\mathbf{v}_T|ds. \tag{2.8}$$

It is straightforward to check that φ is a proper convex function and is weakly lower semicontinuous on V. It is clear that for all $t \in [0,T]$ the domain of $\varphi(t,\cdot)$ contains V, i.e. $D(\varphi(t,\cdot)) \supset V$. The subdifferential of $\varphi(t,\cdot)$ is given by

$$\partial\varphi(t,\mathbf{v}) = \{\mathbf{h} \in V;\ \varphi(t,\mathbf{z}) - \varphi(t,\mathbf{v}) \geq (\mathbf{h},\mathbf{z}-\mathbf{v})_V,\ \ \forall \mathbf{z} \in V\}. \tag{2.9}$$

Using (2.7) and (2.8) there follows

Lemma 2.1 The problem (P_L) is equivalent to

$$\begin{cases} \text{find } \mathbf{w} \in C([0,T];V) \text{ such that } \mathbf{w}(0) = \mathbf{u}_0, \dot{\mathbf{w}} \in L^2(0,T;V) \text{ and} \\ (\mathbf{w}(t) - \mathbf{F}(t),\ \mathbf{v}(t) - \dot{\mathbf{w}}(t))_V + \varphi(t,\mathbf{v}) - \varphi(t,\dot{\mathbf{w}}) \geq 0, \qquad \forall \mathbf{v} \in V, \text{ a.e. on } (0,T). \end{cases} \tag{2.10}$$

Proof. Straightforward.

It follows from (2.10) and (2.9) that we may rewrite the problem as

$$\begin{cases} \text{find } \mathbf{w} \in C([0,T];V) \text{ such that } \mathbf{w}(0) = \mathbf{u}_0,\ \dot{\mathbf{w}} \in L^2(0,T;V) \text{ and} \\ -\mathbf{w}(t) + \mathbf{F}(t) \in \partial\varphi(t,\dot{\mathbf{w}}), \qquad \text{a.e. } t \in (0,T). \end{cases} \tag{2.11}$$

The next result, although simple, is the key to the theorem of this Section. It provides sufficient regularity of φ with respect to t.

Lemma 2.2 There exists a Lipschitz function $k : V \to R_+$ and a function $h \in W^{1,\infty}(0,T)$ such that

$$\varphi(t,\mathbf{v}) + (\mathbf{v},\mathbf{F}(t))_V \leq \varphi(s,\mathbf{v}) + (\mathbf{v},\mathbf{F}(s))_V + k(\mathbf{v})|h(t) - h(s)|, \tag{2.12}$$

$\forall s,t \in [0,T]$.

Proof. First we notice that $\|\mathbf{v}_T\|_{L^4(\Gamma_C)} \leq C_\Gamma \|\mathbf{v}\|_V$ by a trace theorem.
Then define k and h as follows

$$k(\mathbf{v}) = \max\{1, C_\Gamma \|g\|_{W^{1,\infty}}\} \|\mathbf{v}\|_V, \tag{2.13}$$

$$h(t) = \int_0^t [1 + \|\dot{\mathbf{F}}(\tau)\|_V] d\tau. \tag{2.14}$$

It is clear that $k(\mathbf{v})$ is Lipschitz and nonnegative and h continuous. Also $\|g\|_{W^{1,\infty}} = \|g\|_{W^{1,\infty}(0,T;L^{\frac{4}{3}}(\Gamma_C))}$. We claim that with these k and h (2.12) holds. Indeed

$$\begin{aligned} &\varphi(t,\mathbf{v}) + (\mathbf{v},\mathbf{F}(t))_V \leq \varphi(s,\mathbf{v}) + (\mathbf{v},\mathbf{F}(s))_V + |\varphi(t,\mathbf{v}) - \varphi(s,\mathbf{v})| + \\ &+ |(\mathbf{v},\mathbf{F}(t) - \mathbf{F}(s))_V| \leq \varphi(s,\mathbf{v}) + (\mathbf{v},\mathbf{F}(s))_V + \int_{\Gamma_C} |\mathbf{v}_T|\ |g(t,y) - g(s,y)| dy + \\ &+ \|\mathbf{v}\|_V \|\mathbf{F}(t) - \mathbf{F}(s)\|_V \leq \varphi(s,\mathbf{v}) + (\mathbf{v},\mathbf{F}(s))_V + \|\mathbf{v}_T\|_{L^4(\Gamma_c)} \|g\|_{W^{1,\infty}} |t-s| + \\ &+ \|\mathbf{v}\|_V \int_s^t \|\dot{\mathbf{F}}(\tau)\|_V d\tau \leq \varphi(s,\mathbf{v}) + (\mathbf{v},\mathbf{F}(s))_V + k(\mathbf{v})|h(t) - h(s)|, \end{aligned}$$

where (2.13) and (2.14) were used in the last inequality. We are in a position to state

Theorem 2.3 Assume that the compatibility condition

$$\sup_{\mathbf{v}\in V} \{(\mathbf{F}(0) - \mathbf{u}_0, \mathbf{v})_V - \varphi(0, \mathbf{v})\} < +\infty, \tag{2.15}$$

is satisfied. Then problem (P_L), (2.10), has a unique solution.

Proof. We follow the approach of Brezis [B1.pp116-124]. Let $\varphi^*(t,\cdot)$ be the conjugate function of $\varphi(t,\cdot)$, i.e.

$$\varphi^*(t,\mathbf{v}) = \sup_{\mathbf{z}\in V} \{(\mathbf{z},\mathbf{v})_V - \varphi(t,\mathbf{z})\}, \qquad \mathbf{v} \in V, t \in [0,T].$$

Then $\varphi^*(t,\cdot)$ is a convex weakly lower semicontinuous function and for each $t \in [0,T]$, $D(\varphi^*(t,\cdot)) \neq \emptyset$, i.e. it is proper. The last assertion follows from the fact that $\varphi(t,\mathbf{v}) \geq 0\ \forall \mathbf{v} \in V$ hence $\varphi^*(t,\mathbf{0}) = 0, \forall t \in [0,T]$. We rewrite the problem (P_L), (2.11), in the equivalent form

$$\dot{\mathbf{w}} \in \partial\varphi^*(t, \mathbf{F}(t) - \mathbf{w}(t)). \tag{2.16}$$

Recall that $\dot{\mathbf{w}} = d\mathbf{w}/dt$. Let

$$\varphi_1(t,\mathbf{v}) = \varphi(t,\mathbf{v}) - (\mathbf{v}, \mathbf{F}(t))_V$$

and then it follows from Lemma 2.2 that

$$\varphi_1(t,\mathbf{v}) \leq \varphi_1(s,\mathbf{v}) + k(\mathbf{v})|h(t) - h(s)|, \quad \forall t,s \in [0,T], \tag{2.17}$$

where k and h are given by (2.13) and (2.14) respectively. We rewrite (2.16) as

$$-\frac{d}{dt}(-\mathbf{w}) \in \partial\varphi_1^*(t, -\mathbf{w}(t)), \quad \mathbf{w}(0) = \mathbf{u}_0 \ \in V. \tag{2.18}$$

which is equivalent to (2.10).

Now we use Péralba's theorem [Pe2, p93] which guarantees the existence of a unique solution $-\mathbf{w} \in \mathbf{H}^1(0,T;V)$ to (2.18) provided φ_1 satisfies (2.17) and that $-\mathbf{u}_0 \in D(\varphi_1^*(0,\cdot))$. The last condition is exactly the compatibility condition (2.15) and this completes the proof.

Remark 2.4 We could have used the results of Attouch et al [ABDP], or Attouch and Damlamian [AD1, 2] to prove the theorem. Each of these papers deals with the questions of solvability of differential inclusions of the type (2.18). We choose to follow Péralba since his results are used below in the *a* priori estimates that we obtain.

Remark 2.5 It turns out that the compatibility conditions (2.15) have a clear physical interpretation. Indeed if we assume that $\mathbf{u}_0$ is sufficiently smooth then it is straightforward to check that $\mathbf{u}_0$ satisfies (2.15) iff it solves the problem

$$\operatorname{div}\sigma(\mathbf{u}_0) + \mathbf{f}(0) = 0, \qquad \text{in } \Omega,$$

$$\mathbf{u}_0 = 0, \qquad \text{on } \Gamma_U,$$

$$\sigma_N(\mathbf{u}_0) = F_N, \qquad \text{on } \Gamma_C,$$

$$|\sigma_T(\mathbf{u}_0)| \le g(\mathbf{x}, 0), \qquad \text{on } \Gamma_C.$$

Thus the condition (2.15) just states that the system is initially in equilibrium. But the main assumption in a quasistatic problem is that the system is in equilibrium at all times and hence (2.15) is a rather obvious requirement. Stated differently if (2.15) does not hold then the evolution of the system is governed by the full dynamic problem and the assumption that the process is quasistatic is inapplicable.

Remark 2.6 Note that $\varphi_1^*(t, \mathbf{v}) = 0$ for $\mathbf{v} \in D(\varphi_1^*(t, \cdot))$ and hence $\varphi_1^*(t, \cdot)$ is the indicator function of a set.

We turn to obtain some precise *a* priori estimates on the solution $\mathbf{w}$ to (2.10) in terms of g and F_N. In these estimates the inequality

$$C_a \|\mathbf{v}\|^2_{\mathbf{H}^1(\Omega)} \le a(\mathbf{v}, \mathbf{v}) = \|\mathbf{v}\|^2_V \le \frac{1}{C_a} \|\mathbf{v}\|^2_{\mathbf{H}^1(\Omega)}, \quad \forall \mathbf{v} \in V. \tag{2.19}$$

plays an important role and C_a is related to Korn's constant and depends on Ω and on the elastic coefficients $a_{ijk\ell}$. Our first estimate is

Lemma 2.7 Under the assumptions of Theorem 2.3 there holds

$$\|\mathbf{w}\|^2_{C([0,T];V)} \le C\left\{ \|\mathbf{u}_0\|^2_V + \|F_N\|^2_{W^{1,\infty}(0,T;L^{\frac{4}{3}}(\Gamma_C))} + \|f\|^2_{W^{1,\infty}(0,T;\mathbf{L}^2(\Omega))} \right\}. \tag{2.20}$$

We may choose $C = 15(1 + C_a^{-1} + C_\Gamma^2)$.

Proof. Let $\mathbf{v} = 0$ in (2.1) and then let $\mathbf{v} = 2\dot{\mathbf{w}}$, adding we obtain

$$a(\mathbf{w}, \dot{\mathbf{w}}) + \int_{\Gamma_C} g|\dot{\mathbf{w}}_T|ds - \int_{\Gamma_C} F_N \dot{w}_N ds = \int_\Omega \mathbf{f}\dot{\mathbf{w}}dx,$$

which implies

$$\frac{1}{2}\frac{d}{dt}\|\mathbf{w}(t)\|_V^2 + \varphi(t, \dot{\mathbf{w}}_T) = \frac{d}{dt}(\mathbf{F}(t), \mathbf{w}(t))_V - (\dot{\mathbf{F}}(t), \mathbf{w}(t))_V,$$

or

$$\begin{aligned} \frac{1}{2}\|\mathbf{w}(t)\|_V^2 &\le \frac{1}{2}\|\mathbf{u}_0\|_V^2 + (\mathbf{F}(t), \mathbf{w}(t))_V - (\mathbf{F}(0), \mathbf{u}_0)_V - \\ - \int_0^t (\dot{\mathbf{F}}(\tau), \mathbf{w}(\tau))_V d\tau &\le \frac{1}{2}\|\mathbf{u}_0\|_V^2 + 2\|\mathbf{F}(t)\|_V^2 + \frac{1}{8}\|\mathbf{w}(t)\|_V^2 + \frac{1}{2}\|\mathbf{F}(0)\|_V^2 + \\ &+ \frac{1}{2}\|\mathbf{u}_0\|_V^2 + 2\|\dot{\mathbf{F}}\|_{L^1(0,T;V)}^2. \end{aligned}$$

Hence

$$\|\mathbf{w}\|_{C([0,T];V)}^2 \le 10\left\{\|\mathbf{u}_0\|_V^2 + \|\mathbf{F}\|_{L^\infty(0,T;V)}^2 + \|\dot{\mathbf{F}}\|_{L^1(0,T;V)}\right\}. \tag{2.21}$$

But

$$\|\mathbf{F}\|_{W^{1,\infty}(0,T;V)}^2 \le \frac{3}{2}(C_a^{-1} + C_\Gamma^2)\left\{\|\mathbf{f}\|_{W^{1,\infty}(0,T;\mathbf{L}^2(\Omega))} + \|F_N\|_{W^{1,\infty}(0,T;L^{\frac{4}{3}}(\Gamma_C))}\right\},$$

and (2.20) follows when we insert this estimate in (2.21). We turn to an estimate for the time derivative.

Lemma 2.8 Under the assumptions of Theorem 2.3 there holds

$$\begin{aligned} &\|\dot{\mathbf{w}}\|_{L^2(0,T;V)} \le (1 + C_\Gamma\|g\|_{W^{1,\infty}(0,T;L^{\frac{4}{3}}(\Gamma_C))})\cdot \\ &\left\{\sqrt{T} + (C_a^{-1} + C_\Gamma)(\|\mathbf{f}\|_{W^{1,\infty}(0,T;\mathbf{L}^2(\Omega))} + \|F_N\|_{H^1(0,T;L^{\frac{4}{3}}(\Gamma_C))})\right\}. \end{aligned} \tag{2.22}$$

Proof. We use estimates from Péralba [Pe1, Pe2] where he considered the initial value problem

$$-\frac{d\mathbf{z}}{dt} \in \partial\varphi_1^*(t, \mathbf{z}), \qquad \text{a.e. } t \in (0, T) \text{ and } \mathbf{z}(0) = \mathbf{z}_0 \in V. \tag{2.23}$$

Under the assumptions (2.17) and $\mathbf{z}_0 \in D(\varphi_1^*(0,\cdot))$ he proved the existence of the unique solution of (2.23), as we mentioned above. Moreover in [Pel] he obtained an explicit estimate for the time derivative of $\mathbf{z}$ ([Pel, pp18-24] and this estimate can be inferred from the sketch of the proof in [Pe2]). For k and h as above ((2.13) and (2.14) respectively) Péralba's estimate may be written as

$$\begin{aligned}\|\dot{\mathbf{z}}\|_{L^2(0,T;V)} \;\le\; & \frac{1}{2}\ell\|\dot{h}\|_{L^2(0,T)} + \Big\{\sqrt{T}k(0)\|\dot{h}\|_{L^2(0,T)} + \\ & + \frac{1}{4}\ell^2\|\dot{h}\|^2_{L^2(0,T)} + \sup_{0\le s\le T}\varphi_1(s,0) + \varphi_1^*(0,\mathbf{z}_0)\Big\}^{\frac{1}{2}},\end{aligned} \tag{2.24}$$

where ℓ is the Lipschitz constant of k. This estimate is slightly simplified since we set $\mathbf{v}_0 = 0 \in V$ in his original one. In our case $\varphi_1(s,0) = 0 \;\; \forall\, s$ and furthermore $k(0) = 0$. Also

$$\ell = 1 + C_\Gamma \|g\|_{W^{1,\infty}(0,T;L^{\frac{4}{3}}(\Gamma_C))}.$$

Next we have from (2.14) that

$$\begin{aligned}\|\dot{h}\|_{L^2(0,T)} &\le \sqrt{T} + \|\dot{\mathbf{F}}\|_{L^2(0,T;V)} \\ &\le \sqrt{T} + C_a^{-1}\|\dot{\mathbf{f}}\|_{L^2(0,T;\mathbf{L}^2(\Omega))} + C_\Gamma\|\dot{F}_N\|_{L^2(0,T);L^{\frac{4}{3}}(\Gamma_c))}.\end{aligned}$$

Finally $\varphi_1^*(0,\mathbf{z}_0) = 0$ since $\mathbf{z}_0 = -\mathbf{u}_0 \in D(\varphi_1^*(0,\cdot))$ as a consequence of the compatibility condition (2.15). Substituting these inequalities in (2.24) leads to (2.22).

Let $X = W^{1,\infty}(0,T;\mathbf{L}^2(\Omega)) \times W^{1,\infty}(0,T;L^{\frac{4}{3}}(\Gamma_C)) \times W^{1,\infty}(0,T;V)$ then we have

Theorem 2.9 If $\{\mathbf{f}, F_N, \mathbf{u}_0\} \in K \subset X$ where K is a bounded set of X and if for each triplet the compatibility condition (2.15) holds then the family of the corresponding solutions $\{\mathbf{w}\}$ is bounded in X and is precompact in $C([0,T];\mathbf{L}^2(\Omega))$.

Proof. The boundedness follows from Lemmas 2.7 and 2.8. The precompactness of $\{\mathbf{w}\}$ follows from the result of Simon [S].

§3. Existence for the full problem

We use the estimates of the previous Section to prove the existence of a solution to the full problem (1.14) via the Schauder theorem.

We begin with estimates of g_N and g_T, (1.10) and (1.11) respectively. Let $\mathbf{v} \in L^\infty(0,T;\mathbf{L}^q(\Gamma_C))$ for $1 \le q < 4(n \le 3)$. Then we have

$$g_N(\mathbf{v}) = \int_0^t C_N(\mathbf{x})(v_N - g_*)_+^{m_N} \rho(t-\tau)d\tau + C_N(\mathbf{x})(u_N(\tau) - g_*)_+^{m_N}[1 - \int_0^t \rho(t-\tau)d\tau], \tag{3.1}$$

and

$$g_T(\mathbf{v}) = \int_0^t C_T(\mathbf{x})(v_N - g_*)_+^{m_T} \rho(t-\tau)d\tau + C_T(\mathbf{x})(u_N(\tau) - g_*)_+^{m_T}[1 - \int_0^t \rho(t-\tau)d\tau], \tag{3.2}$$

where $\rho = \rho_\epsilon(\epsilon > 0)$ is given in (1.9).

We introduce the notation

$$a_\alpha = 2^{m_\alpha - \frac{3}{4}}, \tag{3.3}$$

$$q_\alpha = \frac{3}{4} - \frac{m_\alpha}{2}, \tag{3.4}$$

$$D_\alpha = \|v_N\|_{L^q(\Gamma_C)}^{m_\alpha} + 2\|g_*\|_{L^q(\Gamma_C)}^{m_\alpha} + \|u_N(0)\|_L^{m_\alpha} q(\Gamma_C), \tag{3.5}$$

where $\alpha = N$ or T. Also c_{imb} is the imbedding constant $V \to L^q(\Gamma_C)$ i.e.

$$\|\mathbf{z}\|_{L^q(\Gamma_C)} \le c_{imb}\|\mathbf{z}\|_V, \qquad \mathbf{z} \in V,$$

where $1 \le q \le \frac{n}{n-2}$ and V is given in (1.13).

We have the estimates

Lemma 3.1 Assume that $\mathbf{v} \in L^\infty(0,T;\mathbf{L}^q(\Gamma_C))$, $\frac{4}{3} \le q < 4$, that $C_N, C_T \in L^\infty(\Gamma_C)$, $C_N, C_T \ge 0$, and let $g_* \in L^\infty(\Gamma_C)$. If $q \ge \frac{4}{3}\max(m_N, m_T)$ then there holds

$$\|g_N(\mathbf{v})\|_{L^\infty(0,T;L^{\frac{4}{3}}(\Gamma_C))} \le a_N \|C_N\|_{L^\infty(\Gamma_C)} |\Gamma_C|^{q_N} D_N, \tag{3.6}$$

$$\|g_T(\mathbf{v})\|_{L^\infty(0,T;L^{\frac{4}{3}}(\Gamma_C))} \le a_T\|C_T\|_{L^\infty(\Gamma_C)}|\Gamma_C|^{q_T}D_T, \tag{3.7}$$

$$\|\dot{g}_N(\mathbf{v})\|_{L^\infty(0,T;L^{\frac{4}{3}}(\Gamma_C))} \le a_N\|\dot{\rho}\|_{L^1(0,T)}\|C_N\|_{L^\infty(\Gamma_C)}|\Gamma_C|^{q_N}D_N, \tag{3.8}$$

and

$$\|\dot{g}_T(\mathbf{v})\|_{L^\infty(0,T;L^{\frac{4}{3}}(\Gamma_C))} \le a_T\|\dot{\rho}\|_{L^1(0,T)}\|C_T\|_{L^\infty(\Gamma_C)}|\Gamma_C|^{q_T}D_T, \tag{3.9}$$

where a_α, q_α and $D_\alpha, \alpha = N$ or T are given above and $|\Gamma_C| = \text{meas}(\Gamma_C)$.

Proof. There holds by (3.1) and the assumptions above

$$\begin{aligned}\|g_N(\mathbf{v})\|_{L^\infty(0,T;L^{\frac{4}{3}}(\Gamma_C))} &\le \|C_N\|_{L^\infty(\Gamma_C)}\|\int_0^T \rho(t-\tau)d\tau\|_{L^\infty(0,T)}\cdot \\ \cdot\|(v_N - g_*)_+^{m_N}\|_{L^\infty(0,T;L^{\frac{4}{3}}(\Gamma_C))} &+ \|C_N\|_{L^\infty(\Gamma_C)}\|(u_N(0)-g_*)_+^{m_N}\|_{L^{\frac{4}{3}}(\Gamma_C)} \\ &\le \|C_N\|_{L^\infty(\Gamma_C)}2^{m_N-\frac{3}{4}}[\|v_N\|^{m_N}_{L^{4m_N/3}(\Gamma_C)} \\ &+2\|g_*\|^{m_N}_{L^{4m_N/3}(\Gamma_C)} + \|u_N(0)\|^{m_N}_{L^{\frac{4}{3}}(\Gamma_C)}].\end{aligned} \tag{3.10}$$

Using Hölder's inequality we obtain

$$\|v_N\|_{L^{4m_N/3}(\Gamma_C)} \le |\Gamma_C|^{q_N}\|v_N\|_{L^q(\Gamma_C)},$$

and

$$\|v_N\|_{L^{4m_T/3}(\Gamma_C)} \le |\Gamma_C|^{q_N}\|v_N\|_{L^q(\Gamma_C)},$$

Substituting these inequalities in (3.10) leads to (3.6). The other estimates are proven in a similar way.

Remark 3.2 The regularization of g_N and g_T in time, i.e. the use of the mollifier ρ, makes the estimates (3.8) and (3.9) possible and this is the mathematical reason for introducing it.

We turn to the definition of the operator G. Assume that $\mathbf{v} \in C([0,T];V)$, $1 \le q < 4; \mathbf{v}(0) = \mathbf{u}_0 \in V$, that $\mathbf{f} \in W^{1,\infty}(0,T;\mathbf{L}^2(\Omega))$ and the compatibility condition

$$\begin{aligned}\sup_{\mathbf{z}\in V}\Big\{ \int_\Omega \mathbf{f}(0)\mathbf{z}dx &- \int_{\Gamma_C} g_N(\mathbf{u}_0)z_N ds - a(\mathbf{u}_0,\mathbf{z}) \\ &- \int_{\Gamma_C} g_T(\mathbf{u}_0)|\mathbf{z}_T|ds\Big\} < \infty,\end{aligned} \tag{3.11}$$

holds true. Let $\mathbf{w}$ be the unique solution to the problem

$$\begin{cases} \text{find } \mathbf{w} \in C([0,T];V) \text{ such that } \mathbf{w}(0) = \mathbf{u}_0, \dot{\mathbf{w}} \in L^2(0,T;V) \text{ and} \\ a(\mathbf{w}, \mathbf{y} - \dot{\mathbf{w}}) + \int_{\Gamma_C} g_T(\mathbf{v})(|\mathbf{y}_T| - |\dot{\mathbf{w}}_T|)ds + \int_{\Gamma_C} g_N(\mathbf{v})(\mathbf{y} - \dot{\mathbf{w}})ds \\ \qquad \geq \int_\Omega \mathbf{f}(\mathbf{y} - \dot{\mathbf{w}})dx, \qquad \forall \mathbf{y} \in V, \text{ a.e. on } (0,T) \end{cases} \tag{3.12}$$

guaranteed by Theorem 2.3 in view of the assumptions above. We define the operator $G : C([0,T];V) \to C([0,T];V)$ by

$$G(\mathbf{v}) = \mathbf{w}. \tag{3.13}$$

It follows from the discussion above that G is well defined. Let $\gamma : C([0,T];V) \to C([0,T];\mathbf{L}^q(\Gamma_C))$ be the trace operator. Then it follows from (3.13) and the structure of our problem that G may be considered as an operator $\hat{G} : C([0,T];\mathbf{L}^q(\Gamma_C)) \to C([0,T];\mathbf{L}^q(\Gamma_C))$ as

$$\hat{G}(\mathbf{y}) = \gamma\mathbf{w}, \qquad \mathbf{y} \in C([0,T];\mathbf{L}^q(\Gamma_C)). \tag{3.14}$$

Clearly $\hat{G}$ is well defined since $\mathbf{w} \in C([0,T];V)$. Below we do not distinguish between these two versions of G. For convenience set

$$U_q = C([0,T];\mathbf{L}^q(\Gamma_C)), \tag{3.15}$$

and define the set K by

$$K = \{\mathbf{z} \in U_q \; ; \; \mathbf{z}(0) = \gamma\mathbf{u}_0\}$$

where $\mathbf{u}_0$ is the initial condition in (3.12). Then K is a nonempty convex closed subset of U_q and $G : K \to K$.

Lemma 3.3 Let $Q = \|\mathbf{v}\|_{U_q}$. Then we have

$$\|\gamma\mathbf{w}\|_{U_q} \leq c_{imb}A(1 + b\, Q^{m_N}), \tag{3.16}$$

where

$$A^2 = 15(1 + C_a^{-1} + C_\Gamma^2)(1 + \|\dot{\rho}\|_{L^1(0,T)})(\|\mathbf{u}_0\|_V^2 + \|\mathbf{f}\|_{W^{1,\infty}(0,T;\mathbf{L}^2(\Omega))} + 2\alpha^2\|g_*\|_{L^q}^{2m_N}), \tag{3.17}$$

$$b = 30\alpha^2(1 + C_a^{-1} + C_\Gamma^2)/A, \tag{3.18}$$

and

$$\alpha = a_N\|C_N\|_{L^\infty}\,(1 + \|\dot{\rho}\|_{L^1}^2)|\Gamma_C|^{q_N}. \tag{3.19}$$

Proof. We already have from (2.20) that

$$\|\mathbf{w}\|_{C([0,T];V)}^2 \leq 15(1 + C_a^{-1} + C_\Gamma^2)\left\{\|\mathbf{u}_0\|_V^2 + \|\mathbf{f}\|_{W^{1,\infty}(0,T;\mathbf{L}^2(\Omega))} + \|g_N(\mathbf{v})\|_{W^{1,\infty}(0,T;L^{\frac{4}{3}}(\Gamma_C))}\right\},$$

and when we insert in the estimates (3.6) and (3.8) we obtain

$$\leq 15(1 + C_a^{-1} + C_\Gamma^2)\left\{\|\mathbf{u}_0\|_v^2 + \|\mathbf{f}\|_{W^{1,\infty}}^2 + \alpha^2(\|v_N\|_{L^q}^{m_N} + \|g_*\|_{L^q}^{m_N})^2\right\},$$

where we used (3.19). Then

$$\leq 15(1 + C_a^{-1} + C_\Gamma^2)\left\{\|\mathbf{u}_0\|_V^2 + \|\mathbf{f}\|_{W^{1,\infty}}^2 + 2\alpha^2\|g_*\|_{L^q}^{2m_N} + 2\alpha^2 Q^{2m_N}\right\}.$$

Using (3.17) and (3.18) we obtain $\|\mathbf{w}\|_V \leq A(1 + bQ^{m_N})$ and using the trace γ and the imbedding constant c_{imb} the inequality (3.16) follows. Next we have

Lemma 3.4 Assume that

$$b(3A)^{m_N} < 1, \tag{3.20}$$

and

$$\|\mathbf{v}\|_{U_q} \leq 3A. \tag{3.21}$$

Then

$$\|G^i(\mathbf{v})\|_{U_q} \leq 3A, \qquad i = 0, 1, 2, \ldots \tag{3.22}$$

where

$$G^0(\mathbf{v}) = G(\mathbf{v}) \text{ and } G^i(\mathbf{v}) = G(G^{i-1}(\mathbf{v})).$$

Proof. Very similar to the proofs of Theorem 2.1 and Lemmas 2.1 and 2.2 in [KMS2].

Remark 3.5 Let us consider the condition (3.20) on the data in more detail. From (3.17)-(3.19) it follows that we may write it as

$$b(3A)^{m_N} = (30 \cdot 3^{m_N})(1 + C_a^{-1} + C_\Gamma^2)\alpha^2 A^{m_N - 1}.$$

Obviously (3.20) holds if either α, i.e. $\|C_N\|_{L^\infty(\Gamma_C)}$, or A, i.e. the initial data $\mathbf{u}_0$, the boundary data g_* and the forces $\mathbf{f}$, are sufficiently small. The assumption (3.16) is stronger than the assumption in Theorem 2.1 of [KMS2].

Formally, Signorini's contact condition is obtained when $C_N \to \infty$. Clearly the approach above rules out the possibility to take this limit.

The lemma implies that the ball B_{3A} of radius $3A$ and center 0 in K is invariant under G. Moreover for each $\mathbf{v} \in B_{3A}$ the orbit of $\mathbf{v}$ under G, i.e. $\{\mathbf{v}^i\}_{i=0}^{\infty}$, $\mathbf{v}^0 = \mathbf{v}$ and $\mathbf{v}^{i+1} = G(\mathbf{v}^i)$, belongs to K and is relatively compact (as we show below). Stated more precisely,

Lemma 3.6 Let $R = 3A$, let the assumptions of Lemma 3.4 hold true and define the set K_R by

$$K_R = \left\{\mathbf{z} \in C([0,T]; \mathbf{L}^q(\Gamma_C));\ \mathbf{z}(0) = \gamma \mathbf{u}_0,\ \|\mathbf{z}\|_{U_q} \le R\right\}. \tag{3.23}$$

Then

$$G(K_R) \subset K_R$$

i.e. K_R is invariant under G.

Proof. Follows from Lemma 3.4.

Clearly $K_R \neq \emptyset$ if $\|\gamma \mathbf{u}_0\|_{L^q(\Gamma_C)} \le 3A$.

We turn to estimate the derivative of $G^i(\mathbf{v})$.

Lemma 3.7 Suppose that the assumptions of Lemma 3.4 are satisfied. Let $\mathbf{v} \in C([0,T];V)$ and define $\mathbf{v}^0 = \mathbf{v}, \mathbf{v}^{i+1} = G(\mathbf{v}^i)$, $i \in N$. Then there exists a constant C_D, independent of i, such that

$$\|\dot{\mathbf{v}}^i\|_{L^2(0,T;V)} = \|dG^i(\mathbf{v})/dt\|_{L^2(0,T;V)} \leq C_D, \qquad \forall i \in N. \tag{3.24}$$

Proof. It follows from Lemma 3.4, the estimates (3.8), (3.9) and the assumptions above and (2.22), similarly to (3.22).

Lemma 3.8 Suppose that (3.11), (3.20) hold true and $\|\mathbf{u}_0\|_{L^q(\Gamma_C)} \leq 3A$. Let $\{\mathbf{v}_k\} \subset K_R$ be a sequence such that $\mathbf{v}_k \to \mathbf{v}$ in U_q. Let $\mathbf{u}_k = G(\mathbf{v}_k)$ and $\mathbf{u} = G(\mathbf{v})$. Then

$$\mathbf{u}_k \to \mathbf{u} \qquad \text{weakly in } H^1(0,T;V), \tag{3.25}$$

and

$$\gamma\mathbf{u}_k \to \gamma\mathbf{u} \qquad \text{strongly in } U_q. \tag{3.26}$$

Proof. It follows from Lemmas 3.4 and 3.7 that the sequence $\{\mathbf{u}_k\}$ is bounded in $H^1(0,T;V)$. Therefore there exists a subsequence which is again denoted by $\{\mathbf{u}_k\}$ such that

$$\mathbf{u}_k \to \mathbf{w} \qquad \text{weakly in } H^1(0,T;V), \tag{3.27}$$

and by the result of Simon [S, p85], using imbedding, there holds

$$\gamma\mathbf{u}_k \to \gamma\mathbf{w} \qquad \text{strongly in } U_q. \tag{3.28}$$

Next we show that $\mathbf{w} = G(\mathbf{v})$. We have

$$\begin{aligned} a(\mathbf{u}_k, \varphi - \dot{\mathbf{u}}_k) &+ \int_{\Gamma_C} g_T(\mathbf{v}_k)(|\varphi_T| - |\dot{\mathbf{u}}_{kT}|)ds \\ &+ \int_{\Gamma_C} g_N(\mathbf{v}_k)(\varphi_N - \dot{\mathbf{u}}_{kN})ds \\ &\geq \int_\Omega \mathbf{f}(\varphi - \dot{\mathbf{u}}_k)dx, \qquad \forall\varphi \in V, \text{ a.e. in } (0,T). \end{aligned} \tag{3.29}$$

Here the subsequence $\{\mathbf{v}_k\}$ corresponds to the subsequence $\{\mathbf{u}_k\}$. In order to obtain the limit $k \to \infty$ we multiply (3.29) by $\psi \in H^1(0,T)$ such that $\psi(0) = \psi(T) = 0, \psi \geq 0$ and integrate over $(0,T)$. Using integration by parts we are led to

$$\begin{aligned}\int_0^T \psi a(\mathbf{u}_k, \varphi)dt &+ \frac{1}{2}\int_0^T \dot{\psi} a(\mathbf{u}_k, \mathbf{u}_k)dt \\ &+ \int_0^T \int_{\Gamma_C} \psi g_T(\mathbf{v}_k)(|\varphi_T| - |\dot{\mathbf{u}}_{kT}|)dsdt \\ &+ \int_0^T \int_{\Gamma_C} \psi g_N(\mathbf{v}_k)(\varphi_N - \dot{\mathbf{u}}_{kN})dsdt \\ &\geq \int_0^T \int_\Omega \psi \mathbf{f}(\varphi - \mathbf{u}_k)dxdt,\end{aligned} \tag{3.30}$$

for all $\varphi \in V, \psi \in H_0^1(0,T)$, $\psi \geq 0$.

We consider each term in its turn. It follows from (3.27) that

$$\int_0^T \psi a(\mathbf{u}_k, \varphi)dt \to \int_0^T \psi a(\mathbf{w}, \varphi)dt \qquad \text{as } k \to \infty. \tag{3.31}$$

also

$$\int_0^T \int_{\Gamma_C} \psi g_T(\mathbf{v}_k)|\varphi_T| \, dsdt \to \int_0^T \int_{\Gamma_C} \psi g_T(\mathbf{v})|\varphi_T|dsdt, \tag{3.32}$$

and

$$\liminf_{k\to\infty} \int_0^T \int_{\Gamma_C} \psi g_T(\mathbf{v}_k)|\dot{\mathbf{u}}_{kT}|dsdt \geq$$

$$\geq \int_0^T \int_{\Gamma_c} \psi g_T(\mathbf{v})|\dot{\mathbf{w}}_T|dsdt$$

since $\mathbf{v}_k \to \mathbf{v}$ strongly and by the convexity of the absolute value function.

$$\int_0^T \int_{\Gamma_C} \psi g_N(\mathbf{v}_k)(\varphi_N - \dot{\mathbf{u}}_{kN})dsdt \to \int_0^T \int_{\Gamma_C} \psi g_N(\mathbf{v})(\varphi_N - \dot{\mathbf{w}}_N)dsdt, \tag{3.33}$$

because of the strong convergence $\mathbf{v}_k \to \mathbf{v}$ and (3.27). Next

$$\int_0^T \int_\Omega \psi \mathbf{f}(\varphi - \dot{\mathbf{u}}_k)dxdt \to \int_0^T \int_\Omega \psi \mathbf{f}(\varphi - \dot{\mathbf{w}})dxdt \tag{3.34}$$

because of (3.27). Comparing (3.31) - (3.34) with (3.30) shows that it remains to obtain the limit of $\int_0^T \dot{\psi} a(\mathbf{u}_k, \mathbf{u}_k)dt$. It is enough to prove that

$$\mathbf{u}_k \to \mathbf{w} \qquad \text{strongly in } C([0,T];V). \tag{3.35}$$

We prove that $\{\mathbf{u}_k\}$ is a Cauchy sequence in the Banach space $C([0,T];V)$. First we note that $-\text{div}\sigma(\mathbf{u}_k) = \mathbf{f} \in \mathbf{L}^2(\Omega)$, $t \in [0,T]$ and therefore we may use Green's formula (see e. g. [DL]), also we note that $\sigma_N(\mathbf{u}_k) \in L^{\frac{4}{3}}(\Gamma_C)$ and $\sigma_T(\mathbf{u}_k) \in L^{\frac{4}{3}}(\Gamma_C)$ uniformly in t. Therefore, since div $\sigma(\mathbf{u}_k - \mathbf{u}_\ell) = 0$, we have

$$\begin{aligned} a(\mathbf{u}_k - \mathbf{u}_\ell, \mathbf{u}_k - \mathbf{u}_\ell) = & \int_{\Gamma_C} \sigma_T(\mathbf{u}_k - \mathbf{u}_\ell)(\mathbf{u}_{kT} - \mathbf{u}_{\ell T})ds \\ & + \int_{\Gamma_C} \sigma_N(\mathbf{u}_k - \mathbf{u}_\ell)(u_{kN} - u_{\ell N})ds. \end{aligned} \tag{3.36}$$

But (3.28) implies that

$$u_{kN}(t) - u_{\ell N}(t) \to 0 \qquad \text{in } L^q(\Gamma_C), \qquad \forall t \in [0,T],$$

$$u_{kT}(t) - u_{\ell T}(t) \to 0 \qquad \text{in } L^q(\Gamma_C), \qquad \forall t \in [0,T],$$

Also it follows from the estimates in the proof of Lemma 3.1 that $\sigma_N(\mathbf{v}_k)$ and $\sigma_T(\mathbf{v}_k)$ are bounded uniformly in $L^{\frac{q}{q-1}}(\Gamma_C)$ provided that $m_N, m_T \leq q-1$. Therefore it follows from (3.36) and (2.19) that $\mathbf{u}_k(t) - \mathbf{u}_\ell(t) \to 0$ in V, uniformly in $t \in [0,T]$, when $k, \ell \to \infty$, hence (3.35) is proved. As a direct consequence there holds

$$\frac{1}{2}\int_0^T \dot{\psi} a(\mathbf{u}_k(t), \mathbf{u}_k(t))dt \to -\int_0^T \psi a(\mathbf{w}, \dot{\mathbf{w}})dt. \tag{3.37}$$

Then (3.31) - (3.35) and (3.37) imply that

$$\begin{aligned} \int_0^T \psi \Big\{ a(\mathbf{w}, \varphi - \dot{\mathbf{w}}) + & \int_{\Gamma_C} g_T(\mathbf{v})(|\varphi_T| - |\dot{\mathbf{w}}_T|)ds \\ & + \int_{\Gamma_C} g_N(\mathbf{v})(\varphi_N - \dot{\mathbf{w}}_N)ds \\ & - \int_\Omega \mathbf{f}(\varphi - \dot{\mathbf{w}})dx \Big\} dt \geq 0, \end{aligned}$$

for all $\varphi \in V$, all $\psi \in H_0^1(0,T)$ such that $\psi \geq 0$. This implies that

$$\begin{aligned} a(\mathbf{w}, \varphi - \dot{\mathbf{w}}) &+ \int_{\Gamma_C} g_T(\mathbf{v})(|\varphi_T| - |\dot{\mathbf{w}}_T|)ds \\ &+ \int_{\Gamma_C} g_N(\mathbf{v})(\varphi_N - \dot{\mathbf{w}}_N)ds \\ &\leq \int_\Omega \mathbf{f}(\varphi - \dot{\mathbf{w}})dx, \qquad \forall \varphi \in V, \text{ a.e. in } (0,T). \end{aligned} \tag{3.38}$$

From the unique solvability of (2.1), by Theorem 2.3, it follows that $\mathbf{w}(t) = \mathbf{u}(t)$ a.e. in $t \in (0,T)$, thus $\mathbf{w} = G(\mathbf{v})$. Also the uniqueness, guaranteed by Theorem 2.3 implies that not only the subsequence converges but actually the whole original sequence $\{\mathbf{u}_k\}$ converges weakly in $H^1(0,T;V)$.

Therefore we have

Corollary 3.9. Assume that (3.11), (3.20) hold true and $\|\mathbf{u}_0\|_{L^q(\Gamma_C)} \leq 3A$. Then $G : K_R \to K_R$ is continuous.

These lead us to our main result

Theorem 3.10 Assume that $\mathbf{f} \in W^{1,\infty}(0,T;V), \mathbf{u}_0 \in V, g_* \in L^\infty(\Gamma_C)$ such that (3.11), (3.20) and that $\|\gamma\mathbf{u}_0\|_{V_q} \leq 3A$. In addition assume that $C_N, C_T \in L^\infty(\Gamma_C)$ $C_N, C_T \geq 0$ a.e. on Γ_C, $1 \leq m_N, m_T \leq q-1$. Then the problem

$$(P) \qquad \begin{cases} \text{find } \mathbf{u} \in C([0,T];V) \text{ such that } \mathbf{u}(0) = \mathbf{u}_0, \dot{\mathbf{u}} \in L^2(0,T;V) \text{ and} \\ a(\mathbf{u}, \varphi - \dot{\mathbf{u}}) + \int_{\Gamma_C} g_T(\mathbf{u})(|\varphi_T| - |\dot{\mathbf{u}}_T|)ds + \int_{\Gamma_C} g_N(\mathbf{u})_N(\varphi_N - \dot{\mathbf{u}}_N)ds \\ \qquad \geq \int_\Omega \mathbf{f}(\varphi - \dot{\mathbf{u}})ds, \qquad \forall \varphi \in V, \text{a.e. } t \in (0,T), \end{cases} \tag{3.39}$$

has at least one solution.

Proof. It is enough to show that the nonlinear operator G defined in (3.13) has a fixed point in K_R. The condition $\|\mathbf{u}_0\|_{L^q(\Gamma_C)} \leq 3A = R$ guarantees that K_R is not empty. Moreover K_R is a closed convex set in U_q. $G(K_R) \subset K_R$ by Lemma 3.6 and G is continuous on K_R by Corollary 3.9. It follows from Lemma 3.7 that $\overline{G(K_R)}$ is a bounded set in $H^1(0,T;V)$

and therefore $\overline{G(K_R)}$ is a compact set (using traces) in $C([0,T];\mathbf{L}^q(\Gamma_C))$ by the result of Simon [S, Corollary 4, p85]. Now the theorem follows from Schauder's fixed point theorem.

Remark 3.11 A careful examination of the results above shows that we might equally well let C_N and C_T depend on time. Then the requirement should be

$$C_N, C_T \in W^{1,\infty}(0,T;L^\infty(\Gamma_C)).$$

Under this assumption we only have to replace $\|C_N\|_{L^\infty(\Gamma_C)}$ and $\|C_T\|_{L^\infty(\Gamma_C)}$ in the estimates by $\|C_N\|_{W^{1,\infty}}$ and $\|C_T\|_{W^{1,\infty}}$.

REFERENCES

[A] Andersson L.E. "A quasistatic frictional problem with normal compliance" preprint, Linköping University, LiTH-MAT-R-89-05

[ABDP] Attouch H., Bénilan P., Damlamian A., Picard C. "Equations d'évolution avec condition unilatérale", *C. R. Acad. Sci. Paris* **279** série A, (1974), 607-609,

[AD1] Attouch H., Damlamian A. "Problémes d'évolution dans les Hilberts et applications", *J. Math pures at appl.* **54**, (1975), 53-74.

[AD2] Attouch H., Damlamian A., "Strong solutions for parabolic variational inequalities", *Nonlinear Anal. TMA*, **2**(3), (1978), 329-353.

[B1] Brézis H. "Problémes unilatéraux", *J. Math. pures et appl.* **51**, (1972), 1-168.

[B2] Brézis H. "Un probléme d'évolution avec constraintes unilatérales dépendant du temps", *C. R. Acad. Sci., Paris* **276**, série A, (1972), 310-312.

[B3] Brézis H. *"Opérateurs maximaux monotones et semi-groupes de contraction dans les espaces de Hilbert et équations d'évolution", Lecture Notes* **5**, North Holland, 1972.

[BT] Bowden F. P., Tabor, D. *The Friction and Lubrication of Solids*, Clarendon Press, Oxford, 1950.

[DL] Duvaut G., Lions J. L. *Inequalities in Mechanics and Physics*, Springer, Berlin, 1976.

[EMS] Elliott C.M., Mikelić A., and Shillor M., "Constrained anisotropic elastic materials in unilateral contact with or without friction", to appear in *Nonlin. Anal. TMA.*

[KMS1] Klarbring A., Mikelić A., Shillor M., "Frictional contact problems with normal compliance". *Int. J. Engng. Sci* **26**(8), (1988), 811-832.

[KMS2] Klarbring A., Mikelić A., Shillor M., "On frictional problems with normal compliance", *Nonlinear Anal. TMA*

[KMS3] Klarbring A., Mikelić A., Shillor M., "Duality applied to contact problems with friction", to appear in *Appl. Math. Optim.*

[KMS4] Klarbring A., Mikelić A. and Shillor M., " The rigid punch problem with friction", preprint.

[KO] Kikuchi N., Oden J. T., *Contact Problems in Elasticity: A Study of Variational Inequalities and Finite Element Methods*, SIAM, Philadelphia, 1988.

[MO] Martins J.A.C., Oden J.T., "Existence and uniqueness results for dynamic contact problems with nonlinear normal and friction interface laws", *Nonlinear Anal. TMA* **11**(3), (1987), 407-428.

[OM] Oden J.T., Martins J.A.C., "Models and computational methods for dynamic friction phenomena", *Computation Meth. Appl. Mech. Engng.* **52**, (1985), 527-634.

[OP] Oden J. T., Pires E., "Analysis of contact problems with non-local friction", *TICOM*, Univ. Texas, Austin, 1981, Report 81-11.

[Pa] Panagiotopoulos P.D., *Inequality Problems in Mechanics and Applications*, Birkhauser, Boston, 1985.

[Pe1] Péralba J. C., "Equations d'évolution dans un espace de Hilbert, associees a des operateurs sous-differentiels", Thesis, Univ. Languedoc, 1973.

[Pe2] Péralba J. C., "Un probléme d'évolution relatif a un opérateur sous-différentiel dépendant du temps", *C. R. Acad. Sci. Paris* **275**, série A, (1972), 93-96.

[RMOC] Rabier P, Martins J.A.C., Oden J.T., Campos L. "Existence and local uniqueness of solutions for contact problems in elasticity with nonlinear friction laws", *Int. J. Engng. Sci.* **24**(11), (1986), 1755-1768.

[RO] Rabier P.J., Oden J.T., "Solution to Signorini-like contact problems through interface models - I. Preliminaries and formulation of variational inequality", *Nonlinear ANal. TMA* **11**(12), (1987), 1325-1350. "II. Existence and uniqueness theorems", *Nonlinear Anal. TMA* **12** (1), (1988), 1-17.

[S] Simon J. "Compact sets in the space $L^P(0,T;B)$", *Ann. Mat. pura et applicata*, (IV), Vol. CXLVI, (1987), 65-96.

International Series of Numerical Mathematics, Vol. 101, © 1991 Birkhäuser Verlag Basel

Debonding of Bimodular Plates

Angelo LEONARDI - Franco MACERI - Elio SACCO

Dipartimento di Ingegneria Civile Edile - Universita` di Roma 2

1. Introduction

In the design of structures made of composite materials, the structural elements must be assembled in such a manner that the overall structure retains its integrity while performing its intended function, subject to loads and enviroments.

For polymer matrix fiber reinforced composites, adhesive bonding and mechanical fasteners (bolts and rivets) can be utilized. Adhesive bonding is preferred because of the continuous connection, whereas when drilling holes for bolts or rivets the fibers are cut, the joining occurs at discrete points, and large stress concentrations arise near each hole drilled.

For these reasons, problems related with the joining of fiber reinforced materials by means of adhesive bonds represent a very important research area. A comprehensive exposition of results in this field can be found in [1] and [2].

Aim of this paper is to present a simple model, able to describe the failure of bonded joints between a rigid support and a multi-layer bimodular plate under transverse loads.

The behavior of this plate is analyzed by using the shear deformation model due to Hencky-Mindlin, based on the assumption that straight lines orthogonal to the middle plane before deformation remain straight but not necessarily perpendicular to the middle surface after deformation. The plate is supposed to be made of several orthotropic layers, with material axes arbitrarily oriented.

A very special stress-strain relationship is considered. We refer to some fiber-reinforced materials, expecially those with very soft matrices (for instance, cord-rubber composites) which have different behavior when loaded in tension or in compression. To describe such kind of non-linear constitutive relation, the model usually employed considers the material as linearly elastic with different elastic moduli in tension and in compression, and it is called "bimodulus" or "bimodular". A quite updated exposition with extensive references of the studies on the constitutive equations of bimodular elastic materials can be found in [3].

The bond material is modelled as an elastic continuum with finite strength in tension. This model was proposed in [4], [5] and [6] in order to describe the delamination phenomenon, and in literature is called "unilateral contact model".

The finite element method is used to analyze the debonding of bimodular plates from a rigid support problem and some numerical examples are given.

2. The mathematical model of the plate

Let Ω be an open bounded set of R^2 with regular boundary $\partial\Omega$. We call a *plate with constant thickness* h the body $\mathcal{P}$:

$$\mathcal{P} = \Omega \times]-(h/2), (h/2)[.$$

Let $(0, e_1, e_2, e_3)$ be a Cartesian coordinate frame, the origin of which belongs to the middle plane of $\mathcal{P}$, with e_3 perpendicular to Ω, and let (x_1, x_2, x_3) be the coordinates of a point $x \in \mathcal{P}$. The Hencky-Mindlin plate theory [7], frequently referred to as the first order shear deformation theory (FSDT), received great attention in the analysis of fiber reinforced composite structures because the elastic transverse shear moduli are significantly smaller than the effective moduli in the directions of the fibers. In this theory the displacement field is assumed to be:

(1) $$u = v + x_3 d,$$

being:

(2) $$v = v_1(x_1,x_2)e_1 + v_2(x_1,x_2)e_2 + v_3(x_1,x_2)e_3 ,$$

(3) $$d = d_1(x_1,x_2)e_1 + d_2(x_1,x_2)e_2 .$$

Note that, if $P_i = e_i \otimes e_i$, then $P = P_1 + P_2$ is the orthogonal projector on the (e_1, e_2) plane and therefore $Pd = d$.

The associated and therefore strain tensor is given by:

(4) $$E = \mathrm{Sym}\,(\nabla v + x_3 \nabla d + d \otimes e_3) ,$$

or, taking into account that v and d do not depend upon x_3, by:

(5) $$E = \mathrm{Sym}\,(\nabla v P + x_3 \nabla d P + d \otimes e_3) .$$

Let us define the resultant stress tensors as:

(6) $$N = \int_h T \, dx_3$$ normal and shear stresses,

(7) $$M = \int_h x_3 T \, dx_3$$ flexural and torsional moments,

where T denotes the Cauchy stress tensor. Let us introduce the applied load densities referred to the middle plane Ω of the plate as:

q normal and transverse loads,

c flexural and torsional couples.

Finally, let us define the boundary load densities on the part $\partial_1\Omega$ of $\partial\Omega$ as:

t normal and transverse loads,

m flexural and torsional couples.

The equilibrium equation, written in variational form, is:

(8) $$\int_\Omega [N \cdot \nabla v_v + M \cdot \nabla d_v + N \cdot d_v \otimes \mathbf{e}_3] \, da$$
$$= \int_{\partial_1 \Omega} [t \cdot v_v + m \cdot d_v] \, ds + \int_\Omega [q \cdot v_v + c \cdot d_v] \, da,$$

with v_v and d_v any variation of the displacement field satisfying the condition $v_v = d_v = 0$ on the part $\partial_2\Omega$ of the boundary on which displacements are assigned.

3. The bimodular constitutive equation

We are interested in fiber-reinforced materials which exibit quite different elastic behavior when loaded in tension or in compression. As a first approximation, the stress-strain behaviour of such materials is often represented as being bilinear, with different slopes (elastic properties) depending upon the sign of the fiber-direction strain. Such model, proposed in [8], is referred in literature as "bimodular fiber-governed macroscopic material model", and is found to agree well with experimental results.

We assume the existence of two symmetric fourth-order elasticity tensors: the one ($\mathbf{C}^+$) describes the behaviour of fibers in elongation and the other one ($\mathbf{C}^-$) in shortening. The local stress-strain relation is written in the form:

$$T = \mathbf{C}^* E , \tag{9}$$

where $\mathbf{C}^*$ is the orthotropic tensor satisfying the condition $E_{33}=0 \Leftrightarrow S_{33}=0$ required in plate theories, defined as:

$$\mathbf{C}^* = \mathbf{C}^+ \qquad \text{if } E\cdot e\otimes e > 0 ,$$

$$\mathbf{C}^* = \mathbf{C}^- \qquad \text{if } E\cdot e\otimes e \leq 0 ,$$

with e the unit vector in the fibers direction. Denoting by $e_\perp$ the unit vector such that $e_3 = e \times e_\perp$, and by:

$$Q_1 = e\otimes e \qquad Q_2 = e_\perp\otimes e_\perp \qquad Q_{23} = e_\perp\otimes e_3 \qquad Q_{13} = e\otimes e_3$$

$$Q_{12} = e\otimes e_\perp \qquad Q_{32} = e_3\otimes e_\perp \qquad Q_{31} = e_3\otimes e \qquad Q_{21} = e_\perp\otimes e.$$

$\mathbf{C}^*$ admits for the FSDT the following representation formula:

$$\begin{aligned} \mathbf{C}^* = {} & \alpha_1^* Q_1\otimes Q_1 + \alpha_2^* Q_2\otimes Q_2 + \beta_1^*(Q_1\otimes Q_2 + Q_2\otimes Q_1) \\ & + \chi_1\gamma_1^*(Q_{23}+Q_{32})\otimes(Q_{23}+Q_{32}) + \chi_2\gamma_2^*(Q_{13}+Q_{31})\otimes(Q_{13}+Q_{31}) \\ & + \gamma_3^*(Q_{12}+Q_{21})\otimes (Q_{12}+Q_{21}) , \end{aligned} \tag{10}$$

where α_1^*, α_2^*, α_3^*, β_1^*, γ_1^*, γ_2^*, γ_3^* are the elasticity constants, and χ_1, χ_2 are the shear correction factors.

Since composite plates are generally made of several laminae with different fibers direction, the constitutive equation for the laminate is obtained by performing the integration layer to layer. By this way, after setting

$$\mathbf{A} = \int_h \mathbf{C}^*(x_3)dx_3 , \quad \mathbf{B} = \int_h x_3\mathbf{C}^*(x_3)dx_3 , \quad \mathbf{D} = \int_h x_3^2\mathbf{C}^*(x_3)dx_3 , \tag{11}$$

the constitutive equations in terms of resultant stresses and displacements are become:

$$N = \mathbf{A}\,[\nabla v + d\otimes e_3] + \mathbf{B}\,[\nabla d] , \tag{12}$$

$$M = \mathbf{B}\,[\nabla v + d\otimes e_3] + \mathbf{D}\,[\nabla d] . \tag{13}$$

4. The debonding model

We consider an elastic plate bonded on a rigid support by means of an adhesive glue and subjected to a transverse load system. We make the assumption that only stresses acting on the glue in e_3 direction are allowed for.

When the limit tension on the glue is attained, debonding occurs. We model such phenomenon by considering the adhesive medium as an uniform spring distribution bed with finite tensile strength r^0.

The spring force density $r = r_3 e_3$ is then assumed to be a function of the transverse deflection in the following way:

(14) $$r_3(u_3) = k(u_3)u_3,$$

with

(15) $$k(u_3) = \begin{cases} k^0 & \text{if } u_3 \leq r^0/k^0 \\ 0 & \text{if } u_3 > r^0/k^0 \end{cases},$$

k^0 being the modulus of the spring.

Figure 1 shows the behaviour of the spring reaction and of the strain energy density γ versus deflection.

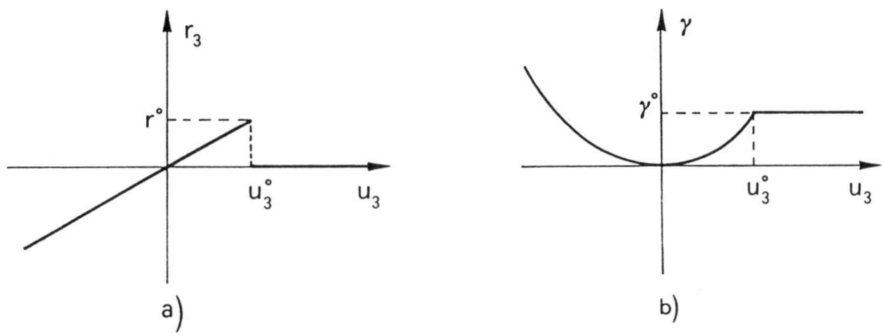

Fig. 1 - Spring reaction (a) and strain energy (b) vs. deflection.

The load q acting on the plate is decomposed into the sum of a known part p, the applied load, and an unknown part r, the spring's reaction, depending upon the deflection u_3:

(16) $$q = p - r = p - r_3 e_3.$$

5. The elastostatic problem

The analysis of a bimodular laminated plate debonding from a rigid support leads to the following non-linear problem: **given**

the field loads (q , c in Ω),

the geometric boundary conditions ($\hat{v}$, $\hat{d}$ on $\partial_2\Omega$),

the natural boundary conditions (t , m on $\partial_1\Omega$),

find the displacement and stress fields satisfying:

the representation formula (1)

the kinematic compatibility equation (4)

the equilibrium equation (8)

the internal bimodular constitutive equations (12) (13)

the external unilateral conditions (15) (16).

It is worth noting that, the strain energy function being not coercive and not convex, existence and uniqueness of solution of this problem cannot be proved in a general case.

Closed-form solutions for this problem are possible only in very special cases. For instance, let us consider a narrow strip of a long rectangular plate made of one orthotropic bimodular layer, resting on a spring bed (with finite tensile strength) and subject to constant displacement δ_3 given on the line $x_1 = x_3 = 0$, as shown in fig. 2 (x_1, x_2, x_3 are also axes of orthotropy).

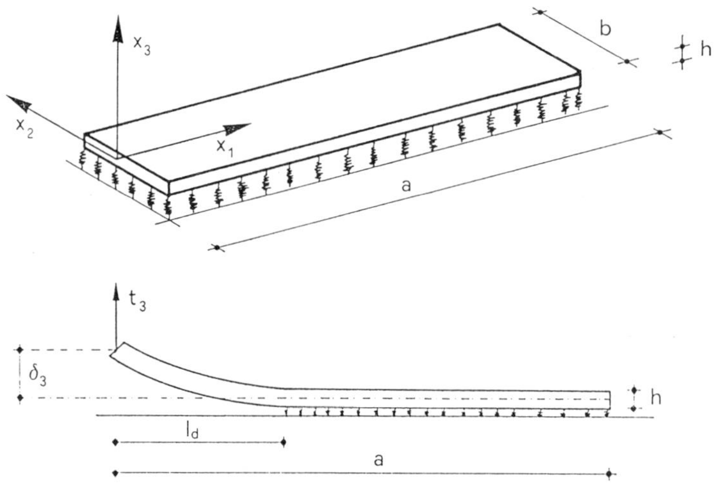

Fig. 2 -Narrow strip plate. Geometry and loading.

By the effect of δ_3, the plate is deformed in cilyndrical bending, and the relations between the debonding length l_d, the edge line reaction t_3 and the transverse displacement δ_3 at $x_1=0$ are:

$$(17) \qquad l_d = \frac{\sqrt{2D\gamma}}{t_3} - \lambda ,$$

$$(18) \qquad \delta_3 = \sqrt{\frac{2\gamma}{D}} \left[\frac{\lambda^2}{2} + \frac{l_d^3}{3(l_d+\lambda)} + \frac{l_d\lambda(2l_d+\lambda)}{2(l_d+\lambda)} \right] ,$$

where D is the flexural rigidity and λ is the characteristic length:

$$(19) \qquad D = \sqrt{\alpha_1^+\alpha_1^-}\, h^3 \frac{3\alpha_1^- - 2\sqrt{\alpha_1^+\alpha_1^-} + 3\alpha_1^+}{12(\sqrt{\alpha_1^+} + \sqrt{\alpha_1^-})} \quad , \quad \lambda = \sqrt[4]{\frac{4D}{k_0}} .$$

We recall that the elasticity constants $\alpha_1{}^*$ are related with the technical constants (E^* Young's moduli and ν^* Poisson's ratios) by the relation:

$$(20) \qquad \alpha_1{}^* = \frac{E_1^*}{1 - \nu_{12}^*\nu_{21}^*} .$$

In [5] it is proved that, as $k\to\infty$ or equivalently $\lambda\to 0$, the unilateral solution converges to the solution given by fracture mechanics, that is:

$$(21) \qquad l_d = \frac{\sqrt{2D\gamma}}{t_3} \quad , \quad \delta_3 = \sqrt{\frac{2\gamma}{D}}\, \frac{l_d^3}{3} .$$

In more general cases, closed-form solutions cannot be obtained. Therefore, it looks interesting to develop a numerical approach.

6. The finite element formulation

Let us denote by Ω_n a polygon with vertices on $\partial\Omega$ such that meas $(\Omega_n - \Omega) << $ meas (Ω) and let us subdivide Ω_n into finite elements Ω_e. The displacement fields v and d on Ω_e are described by:

$$(22) \qquad v = v^i\psi^i \qquad\qquad d = d^i\psi^i ,$$

(summation convention holds) where v^i, d^i are nodal parameters vectors and ψ^i are shape functions.

Substituting (22) into the variational equilibrium equation (8) and taking (12) and (13) into account, we obtain for each element and for any admissible v_v^i and d_v^i:

$$(23)\qquad \int_{\Omega_e} \{ \mathbf{A}[v^j\otimes\nabla\psi^j + Pd^j\otimes e_3\psi^j] + \mathbf{B}[Pd^j\otimes\nabla\psi^j] \}\cdot v_v^i\otimes\nabla\psi^i\, da$$
$$= \int_{\Omega_e} q\cdot v_v^i\psi^i da + \int_{\partial\Omega_e} t_e\cdot v_v^i\psi^i ds ,$$

$$(24)\qquad \int_{\Omega_e} \{ \mathbf{B}[v^j\otimes\nabla\psi^j + Pd^j\otimes e_3\psi^j] + \mathbf{D}[Pd^j\otimes\nabla\psi^j] \}\cdot Pd_v^i\otimes\nabla\psi^i\, da$$
$$+ \int_{\Omega_e} \{ \mathbf{A}[v^j\otimes\nabla\psi^j + Pd^j\otimes e_3\psi^j] + \mathbf{B}[Pd^j\otimes\nabla\psi^j] \}\cdot Pd_v^i\otimes e_3\psi^i\, da$$
$$= \int_{\Omega_e} c\cdot Pd_v^i\psi^i\, da + \int_{\partial\Omega_e} m_e\cdot Pd_v^i\psi^i\, ds ,$$

where t_e and m_e are the boundary interactions between elements.

Setting:

$$A^{ij}=\int_{\Omega_e}\mathbf{A}:\nabla\psi^i\otimes\nabla\psi^j da \qquad \tilde{A}^{ij}=\int_{\Omega_e}\mathbf{A}:\nabla\psi^i\otimes e_3\psi^j da \qquad \hat{A}^{ij}=\int_{\Omega_e}\mathbf{A}:e_3\psi^i\otimes e_3\psi^j da$$

$$B^{ij}=\int_{\Omega_e}\mathbf{B}:\nabla\psi^i\otimes\nabla\psi^j da \qquad \tilde{B}^{ij}=\int_{\Omega_e}\mathbf{B}:\nabla\psi^i\otimes e_3\psi^j da \qquad D^{ij}=\int_{\Omega_e}\mathbf{D}:\nabla\psi^i\otimes\nabla\psi^j da ,$$

and

$$q^i = \int_{\Omega_e} q\psi^i da \qquad c^i = \int_{\Omega_e} c\psi^i da \qquad t_e^i = \int_{\partial\Omega_e} t_e\psi^i ds \qquad m_e^i = \int_{\partial\Omega_e} m_e\psi^i ds ,$$

where the : product is such that $(\mathbf{M}:V)_{ij}=\mathbf{M}_{ikjl}V_{kl}$, equations (23) and (24) can be written in the form:

$$(25)\qquad A^{ij}v^j + [\tilde{A} + B]^{ij}Pd^j = q^i + t_e^i ,$$

$$(26)\qquad P\,[\tilde{A}^T + B]^{ij}v^j + P[\hat{A} + \tilde{B} + \tilde{B}^T + D]^{ij}Pd^j = c^i + m_e^i .$$

Recalling equation (16), the load acting on the plate in the finite element equation (25) is given by:

$$(27)\qquad q^i = p^i - R^{ij}(v)v^j ,$$

being:

$$p^i = \int_{\Omega_e} p\psi^i da \qquad R^{ij}(v) = \int_{\Omega_e} k(v)P_3\psi^i\psi^j\, da .$$

Finally, denoting by K^{ij}, x^j, f^i the stiffness matrix, the unknown vector and the data vector for each element of the mesh:

$$K^{ij} = \begin{bmatrix} A^{ij} + R^{ij}(v) & (\tilde{A} + B)^{ij} P \\ P(\tilde{A}^T + B)^{ij} & P(\hat{A}+\tilde{B}+\tilde{B}^T+D)^{ij} P \end{bmatrix} \quad x^j = \begin{bmatrix} v^j \\ d^j \end{bmatrix} \quad f^i = \begin{bmatrix} p^i + t_e^i \\ c^i + m_e^i \end{bmatrix}$$

equations (23) and (24) can be written in the symbolic form:

(28) $\qquad K^{ij}\, x^j = f^i$.

Performing the assembly of the finite-element equations (28) in the classical way, and recalling that $\mathbf{A}$, $\mathbf{B}$, $\mathbf{D}$ depend on v, d and that R depends on v, we get the non linear problem:

(29) $\qquad K(x)\, x = f.$

7. Numerical procedure and results

In order to solve numerically the non-linear problem (29), the following iterative procedure, based on the well-known *direct iteration (Picard) method* has been used:

Procedure (a)

a-1) set $v=d=0$ in Ω_n and on $\partial_1\Omega_n$ and $v=\hat{v}$, $d=\hat{d}$ on $\partial_2\Omega_n$. Assume $\mathbf{C}^* = \mathbf{C}^+$ in the formula (11);

a-2) construct K and f;

a-3) find x from equation (29);

a-4) once v and d have been determined, compute the fiber elongation zone at the Gauss points of each element and for each layer, and then change the constitutive tensors according to formula (11);

a-5) identify the Gauss points at which the displacement u_3 is greater than u_3^0, and assume $k=0$ for the numerical Gauss integration in these points;

a-6) construct the new stiffness matrix K_n and find $x_n = K_n^{-1} f$

a-7) test the solution accuracy by evaluation of $\|x - x_n\|$. If it is unsatisfactory, come back to step *a-4)* by setting $x = x_n$ and $K = K_n$.

To improve the numerical efficiency of this procedure, that is, to avoid a large number of elements and to describe with a better accuracy the debonding line defined by $u_3 = u_3^0$, we introduce the remeshing of the domain in the following way:

Procedure (b)

b-1) fix an initial debonding line and a mesh with vertices on this line;

b-2) compute x by means of procedure *(a)*, excluding step *a-5)*;

b-3) once v, d are known, compare displacements on the debonding line with the u_3^0 value, in norm; if the result is unsatisfactory, define a new debonding line (by means of interpolation on the displacement field), construct a new mesh with vertices on this line and go to step *b-2)*; otherwise, stop the procedure.

We point out that, to match a regular polygonal mesh, bilinear isoparametric quadrilateral elements were used.

In the following, we present results for some rectangular plates made of one or more layers of orthotropic laminae. The bimodular material used in calculations is the aramid-rubber, the elastic properties of which are:

	tensile	compressive
major Young's modulus (GPa)	3.58420	0.01200
minor Young's modulus (GPa)	0.00909	0.01200
major Poisson's ratio (dimensionless)	0.416	0.205
longitudinal-transverse shear modulus (GPa)	0.00370	0.00370
longitudinal-thickness shear modulus (GPa)	0.00370	0.00370
transverse-thickness shear modulus (GPa)	0.00290	0.00499 .

As a first example, we examined the case of a narrow rectangular plate, where fibers are longitudinal and (Fig. 2)

$$a = 1\text{m} \qquad b = 0.01\text{m} \qquad h = 0.1\text{m} .$$

Two sets of properties for the glue are considered:

$$k^0 = 0.01 \text{ GPa/m}, \quad \gamma = 20 \text{ Pam} \qquad \text{(case a)},$$

or:

$$k^0 = 1 \text{ GPa/m}, \quad \gamma = 2000 \text{ Pam} \qquad \text{(case b)}.$$

The meshing criterion for this case is shown in Fig. 3. The initial position of the debonding line is fixed, and the unglued part is divided into 6÷8 elements; in the glued part a number (from 15 to 20) elements close to debonding line is introduced. The length of the elements is small (1/3÷1/4) if compared with the characteristic length λ (19). Far from the unglued part a coarse mesh is sufficient.

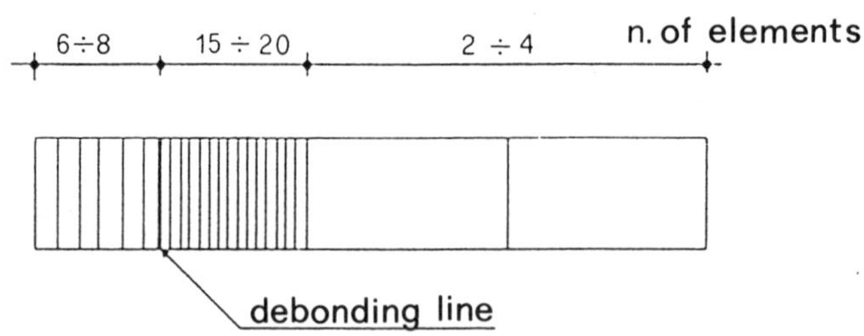

Fig. 3 - FEM mesh for narrow strip plate.

In view of a first evaluation of the results obtained, we made computations by taking very high values of shear moduli, in order to simulate the pure bending behaviour of classical plate theory (CPT), analytically examined in Sect. 5. Computations were made, in this case, by using a reduced integration technique on the shear-type terms, due to the high values taken for the fictitious shear moduli.

In Figures 4 and 5 results for the two cases $k^0 = 1$ and $k^0 = 0.01$ are shown. We put into evidence that the analytical values of l_d and t_3 obtained by (21) agree with those obtained by the numerical procedure. In the same figures, results arising from FSDT theory are shown. In both cases, the full integration rule was applied. The strong influence of shear deformability clearly appears.

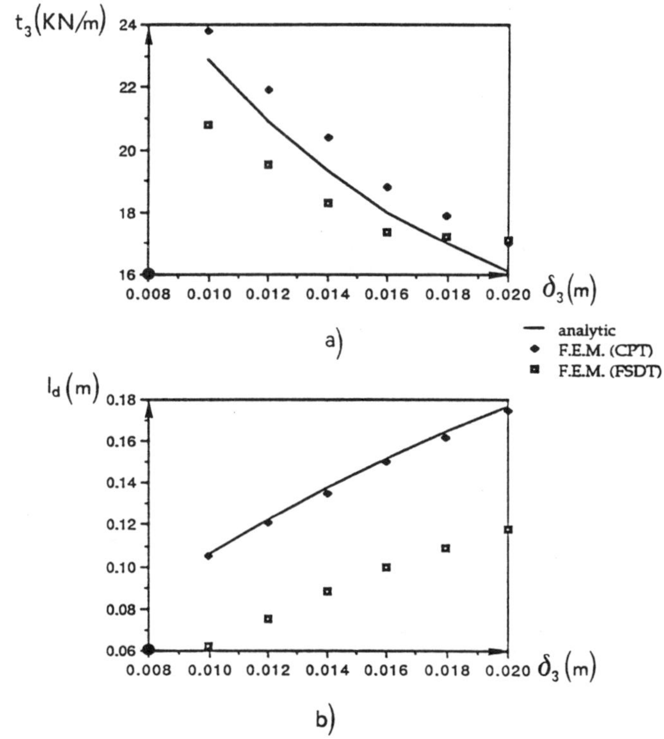

Fig. 4 - Reaction (a) and debonding length (b) vs displacement for narrow strip plate (case a).

As a final remark, we observe that the relationship between the imposed displacement δ_3 and the reaction t_3 is decreasing, that is, the elastostatic problem in terms of prescribed load t_3 on the line $x_1 = x_3 = 0$ is unstable. The prescribed edge displacement approach used in this paper overrides this difficulty and allows us to determine the debonding propagation law.

A second class of examples deals with a square two-layer laminate (a = b = 1m) with thickness h = 0.1m. Three cases are analyzed:

- cross-ply (0/0)
- angle-ply (0/45)
- cross-ply (0/90).

The characteristics of the glue are: $k = 0.01$ GPa/m, $\gamma = 20$ Pam.

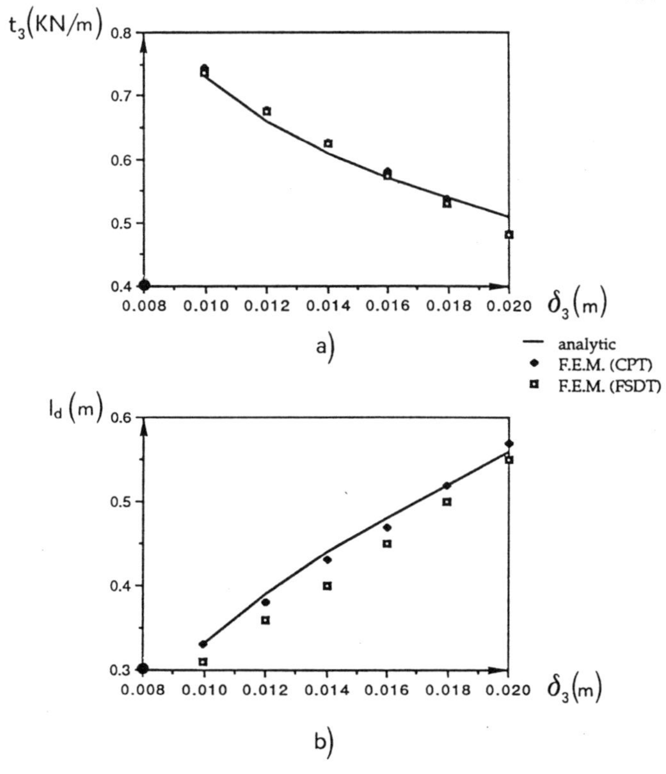

Fig. 5 - Reaction (a) and debonding length (b) vs displacement for narrow strip plate (case b).

A pointwise displacement δ_3 is imposed at a corner, and the external loads are zero.

Concerning the meshing criterium, we define the initial mesh by dividing the square into a given number NS of sectors by equally spaced radii from the loaded corner (Fig. 6). Moreover, we give an initial position of the debonding line, assumed to be circular, of radius R2. The plate area is then divided into four parts. The refined mesh in the glued zone near the debonding line (part 3) is identified by the radius R3 and by the number of elements ND3. Radius R1 identifies both the annular zone

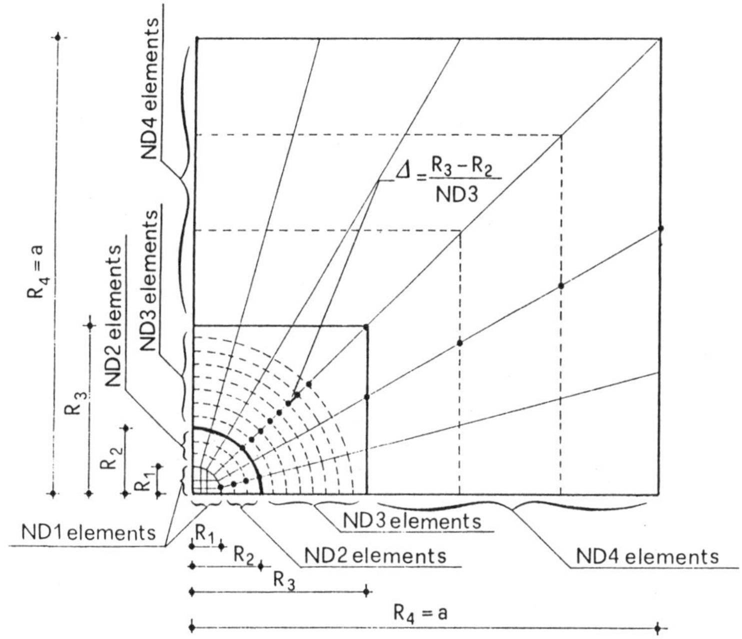

Fig. 6 - Meshing criterion for square plate.

near the debonding line (part 2), where meshing is defined by number of elements ND2, and a circular zone near the loaded corner where the mesh remains fixed during successive remeshing steps (part 1). Between R1 and R2 and between R2 and R3 the mesh nodes are established on circles and radii. Between R3 and R4 = a (part 4) each radial segment is divided into ND4 equal parts in order to define the nodes. For R<R1, nodes belongs to lines parallel to the plate sides and passing through nodes belonging to the circle with R=R1. For practical computations, R1 is given in such a way that area with R<R1 is unglued.

During remeshing, a new debonding line is defined; then, being fixed ND1, ND2, ND3, ND4, we maintain the length R3 - R2 as a constant, that is, the radial length of elements in the glued zone adjacent to debonding line along each radius does not change; the radial lengths of elements in part 2 and part 4 are different radius by radius, because of the fact that, in general, the debonding line is not a circle. By this way, the number of nodes remains the same at each step with a shape variation only of quadrilateral elements.

In Fig. 7 a picture from a computer screen is taken, showing a typical evolution of meshing during iterations. (ND1 = ND2 = 3, ND3 = 8, ND4 = 2). In Tab. 1 debonding radii l_{d_1}, l_{d_2}, l_{d_3} (Fig. 8) versus δ_3 are given. In Fig. 9, the corner reaction T_3 versus δ_3 is represented. For all these cases full integration rule was applied.

Fig. 7 - Meshing evolution during iterations.

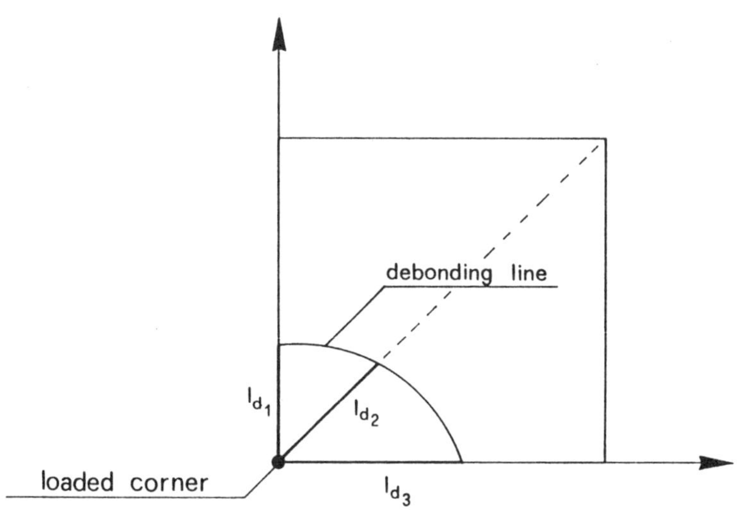

Fig. 8 - Debonding radii in square laminate.

8. Conclusions

The debonding from a rigid support of simple structural elements made of bimodular materials is studied, and an algorithm, based on finite element discretization, is given.

Tab. 1 - Debonding radii (m) vs displacement (m) in square laminates

	cross-ply(0/0)			angle-ply(0/45)			cross-ply(0/90)		
δ_3	l_{d_1}	l_{d_2}	l_{d_3}	l_{d_1}	l_{d_2}	l_{d_3}	l_{d_1}	l_{d_2}	l_{d_3}
0.010	.1742	.1672	.2239	.2068	.1862	.2023	.2284	.1680	.1729
0.012	.2085	.2017	.2697	.2514	.2293	.2459	.2745	.2025	.2070
0.014	.2404	.2339	.3123	.2921	.2691	.2857	.3178	.2347	.2387
0.016	.2712	.2649	.3536	.3309	.3071	.3237	.3595	.2656	.2692
0.018	.3012	.2950	.3939	.3681	.3438	.3603	.4001	..2956	.2989
0.020	.3307	.3246	.4337	.4065	.3815	.3978	.4403	.3251	.3282

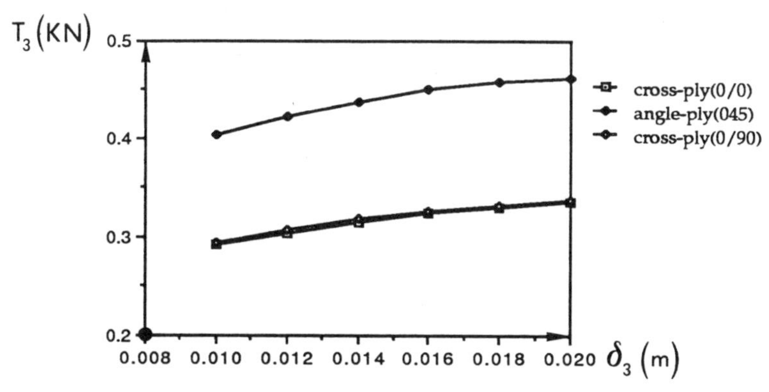

Fig 9 - Reaction vs displacement in square laminate.

Numerical results are given for two model problems concerning a beam-like orthotropic plate and a two-layer laminate.

Results are validated by comparison with analytical solutions available in the simplest case and allow to get a quantitative measure of the debonding phenomena.

Acknowledgement

Financial support of Ministry of Education (MPI) is acknowledged.

References

[1] Baker,A.A. - Jones,R. (Editors) Bonded Repair of Aircraft Structures. (Engineering Application of Fracture Mechanics, Vol. 7) Martinus Nijhoff Publishers, 1988.

[2] Vinson,J.R. - Sierakowski,R.L. The Behavior of Structures Composed of Composite Materials. Martinus Nijhoff Publishers, 1986.

[3] Tabbador,F. A Survey of Constitutive Equations of Bimodulus Elastic Materials. Symposium on "Mechanics of Bimodulus Materials" 1979 Winter Annual Meeting in New York edited by C.W. Bert. December 2-7, 1979.

[4] Frémond,M. Contact Unilatéral avec Adhérence. Proceedings of the Second Meeting on Unilateral Problems in Structural Analysis. CISM Course and Lectures Nº 288, edited by G. Del Piero & F. Maceri. Springer-Verlag 1985.

[5] Grimaldi,A. - Reddy,J.N. On Delamination in Plates: A Unilateral Contact Approach. Proceedings of the Second Meeting on Unilateral Problems in Structural Analysis. CISM Course and Lectures Nº 288, edited by G. Del Piero & F. Maceri. Springer-Verlag 1985.

[6] Ascione,L. - Bruno,D. On the Delamination Problem of Two-Layer Plates. Proceedings of the Second Meeting on Unilateral Problems in Structural Analysis. CISM Course and Lectures Nº 288, edited by G. Del Piero & F. Maceri. Springer-Verlag 1985.

[7] Mindlin,R.D. Influence of Rotatory Inertia and Shear on Flexural Motions of Isotropic, Elastic Plates. J. Appl. Mech., Vol.18, 1951.

[8] Bert, C.W. Models for Fibrous Composites with Different Properties in Tension and Compression. J. Engng. Materials Tech., Trans. ASME, 994(4), 1977.

Dr. Ing. Angelo Leonardi
Prof. Ing. Franco Maceri
Dr. Ing. Elio Sacco
Dipartimento di Ingegneria Civile Edile
Universita' di Roma 2
Via Orazio Raimondo
00173 Roma - Italia

International Series of Numerical Mathematics, Vol. 101, © 1991 Birkhäuser Verlag Basel

REMARKS ON A NUMERICAL METHOD FOR UNILATERAL CONTACT INCLUDING FRICTION

C. LICHT * , E. PRATT ** , M. RAOUS**

1. Introduction

We consider a Signorini problem (unilateral problem) with Coulomb friction in elasticity, under a small strain hypothesis. For proportional loading, the problem is equivalent to the static friction one introduced by Duvaut and Lions ,[72].The general problem has been numerically treated through an incremental formulation in Raous et al ,[88].

For the static problem, Duvaut,[80],and Cocu,[84], give an existence and uniqueness result for small values of the friction coefficient μ and for a regularized problem obtained using a non local definition for the contact forces. Using the discretization of the regularized problem, Jeannin ,[85],shows the convergence of the approximate solution when the discretization step h tends to zero. Using a compliance regularization, allowing a controlled penetration, Oden - Martins,[85],and Klarbring et al,[88],give an existence result and obtain uniqueness under the hypothesis that the friction coefficient is small.

For the initial problem without regularization, Necas et al,[80], give an existence result for small values of μ for a strip. This is done through the introduction of a special Sobolev space for the displacement which gives a compactness property. Jarusek ,[83], extends the result to bounded domains.

A mixed finite element approximation has been studied by Haslinger,[83], a relationship between the continuous problem and its discretization is established.

In this paper, we propose an algorithm concerning a discretized problem associated with the initial one without the regularization of Duvaut and we give a theorem of existence. For small values of μ a fixed point theorem may be used which implies the uniqueness of the solution as well as the convergence of the algorithm. However, the relationship between the continuous problem and the discretization proposed here is still an open problem.

Numerical tests on the convergence illustrate these theoretical results. Special attention is paid to the influence of the friction coefficient and to the dependence on the mesh size. The numerical convergence is better than the theoretical previsions : the conditions required by the fixed point theorem seem to be very strong.

2. The continuous problem

2.1 The mechanical problem for proportional loading

We consider an elastic solid occupying a bounded regular domain Ω of $\mathbb{R}^n$ with a regular boundary $\Gamma = \Gamma_1 \cup \Gamma_2 \cup \Gamma_3$ and submitted to a volume force density $\lambda(t)\ \varphi_1(x)$ in Ω and to a surface force density $\lambda(t)\ \varphi_2(x)$ on Γ_2 (t and x denote respectively the time and the space variables and λ is an increasing positive mapping such that $\lambda(0) = 0$). The solid is clamped on one part of its boundary Γ_1. On Γ_3 the solid is in unilateral contact with friction with a rigid obstacle.

For the quasi-static problem with initial conditions equal to zero, because of the specific nature of the loading, the displacement $U(x,t)$ and the constraint $\Sigma(x,t)$ may be sought as $(U(x,t), \Sigma(x,t)) = \lambda(t)(u(x), \sigma(x))$ where the equations relating u to σ are as follows :

(1) $\sigma_{ij} = K_{ijkh}\ e_{kh}$ in Ω ,

(2) $e_{ij} = \frac{1}{2}\ (u_{i,j} + u_{j,i})$ in Ω ,

(3) $\sigma_{ij,j} = -\ \varphi_{1i}$ in Ω ,

(4) $\sigma_{ij} n_j = \varphi_{2i}$ on Γ_2 ,

(5) $u_i = 0$ on Γ_1 ,

(6) $\sigma_{ij} n_j = F_i$

(7) $u_N \leqslant 0$

(8) $F_N \leqslant 0$

(9) $u_N . F_N = 0$

(10) $|F_T| \leqslant \mu\ |F_N|$

(6)–(10) on Γ_3 ,

with

(11) $|F_T| < \mu\ |F_N| \Rightarrow u_T = 0$,

and

(12) $|F_T| = \mu\ |F_N| \Rightarrow -\ u_T$ is collinear with F_T and with the same direction ,

where :

- K_{ijkh} are the elasticity coefficients ,
- n_j are the components of the external normal to Γ ,
- $u_N = u_i n_i$ and $u_{Ti} = u_i - u_N n_i$,
- $F_N = \sigma_{ij} n_i n_j$ and $F_{Ti} = \sigma_{ij} n_j - F_N n_i$,
- μ is the friction coefficient.

2.2 <u>A mathematical formulation</u>

The following assumptions are set on the data :

(13) $$\begin{cases} K_{ijkh} = K_{jikh} = K_{ijhk} \in L^{\infty}(\Omega) , \\ \exists\ \alpha_0 : \quad K_{ijkh}\ \tau_{kh}\ \tau_{ij} \geqslant \alpha_0\ \tau_{ij}\ \tau_{ij} \qquad \forall\ \tau_{ij} = \tau_{ji} ; \end{cases}$$

(14) $$\mu \in C^1\left(\overline{\Gamma_3}\right) ,$$

$$\exists\ \mu_0 > 0 \quad \text{such that } \mu(x) \geqslant \mu_0 > 0 \qquad \forall\ x \in \overline{\Gamma_3} ,$$

(15) $$\varphi_1 \in L^2(\Omega)^n , \quad \varphi_2 \in L^2(\Gamma_2)^n .$$

We then set :

(16) $$V = \{ v \in H^1(\Omega)^n \ / \ v = 0 \text{ on } \Gamma_1 \} ,$$

(17) $$\mathbb{K} = \{ v \in V \ / \ v_N \leqslant 0 \text{ on } \Gamma_3 \} ,$$

(18) $$a(u,v) = \int_{\Omega} K_{ijkh}\ e_{kh} e_{ij}\ dx ,$$

(19) $$L(v) = \int_{\Omega} \varphi_1 v\ dx + \int_{\Gamma_2} \varphi_2 v\ ds .$$

Let ω be an element of V . If $s = Ke(\omega)$ is such that div $s = -\varphi_1$, we can generalize (on the basis of Green's theorem)the notion of stress vector s.n on Γ_3 by introducing the element $F(\omega) \in H^{-1/2}(\Gamma_3)^n$ defined by :

(20) $$\langle F(\omega), v\rangle = a(\omega,v) - L(v) \qquad \forall\ v \in V \ ;\ v=0 \text{ on } \Gamma_1 \cup \Gamma_2 ,$$

where $\langle .,.\rangle$ is the duality product between $H^{-1/2}(\Gamma_3)^n$ and $H^{1/2}(\Gamma_3)^n$. We shall maintain the same notation, i.e. $\langle .,.\rangle$, for the duality product between $H^{-1/2}(\Gamma_3)$ and $H^{1/2}(\Gamma_3)$.

Because of the regularity of the boundary Γ, we can generalize the notion of normal component $F_N(\omega)$ of the stress vector by

(21) $$\langle F_N(\omega) , \varphi \rangle = \langle F(\omega), \varphi n \rangle \qquad \forall\ \varphi \in H^{1/2}(\Gamma_3) .$$

A weak formulation of the system of local relations (1) to (12) is then the following problem (cf Necas ,[80], Jarusek ,[83]) :

(P) $$\left[\begin{array}{l} \text{find } u \in \mathbb{K} \text{ such that} \\ a(u,v-u) - \langle F_N(u), \mu\ (|v_T| - |u_T|)\ \rangle \geqslant L(v-u) \qquad \forall\ v \in \mathbb{K} . \end{array} \right.$$

By introducing an extra assumption on μ ,Necas, Haslinger and Jarusek have shown (cf Necas et al ,[80], Jarusek,[83]), that this implicit variational inequation (P) has a solution.

2.3 Remark on the contact stresses written in terms of measures

Through classical arguments, equation (3) is shown to be verified in the distribution sense together with the limit conditions (4) (resp.(5)) on Γ_2 (resp. Γ_1) in the trace sense.

By choosing $v=u+\psi$, ψ being regular and with a normal (resp. tangential) component negative (resp. null) on Γ_3, one obtains that F_N belongs to $H^{-1/2}(\Gamma_3)$; this negative distribution is therefore a negative measure, evidently equal to the opposite of its absolute value. Then by choosing once again $v=u+\psi$, ψ being regular but this time with a normal component null on Γ_3 one obtains :

$$\langle F_T , \psi_T \rangle \leqslant \langle \mu \, |F_N| , |\psi_T| \rangle ,$$

which establishes that F_T (and therefore F) is a measure and that condition (10), which expresses the fact that the stress vector along Γ_3 belongs to the Coulomb cone, is satisfed in a measure sense.

3. A finite dimensional problem

To problem (P) is associated a discretised problem by the use of a finite element method .The discrete problem is shown to have a solution which, under a condition on the friction coefficient μ, is unique.

3.1 Formulation of the problem in a finite dimensionnal space U^h

We shall adopt the following notations :

- $\mathcal{T}^h$ is a regular triangulation of the domain Ω : $\overline{\Omega} = \cup_{K\in\mathcal{T}^h} K$ (cf Ciarlet [78]) ,

- Let us set :

$$X^h = \left\{ w \in C^0(\overline{\Omega}) / \; w\Big|_{K\in\mathcal{T}^h} \in P_1 \right\} ,$$

where P_1 is the set of polynomials of degree 1 ,

- $V^h = \left\{ v \in (X^h)^n \;\; / \; v = 0 \text{ on } \Gamma_1 \right\}$,
- $\mathbb{K}^h = \left\{ v \in V^h \;\; / \; v_n \leqslant 0 \text{ on } \Gamma_3 \right\}$,
- $\tilde{X}^h$ is the space of traces of X^h on Γ_3 ,
- $\tilde{V}^h$ is the space of traces of V^h on Γ_3 .

We then set $\{w_i\}$ (resp. $\{v_i\}$), $i = 1 \ldots M$ (resp. P) a basis of X^h (resp. V^h) and $\left\{\tilde{w}_i\right\}$ (resp. $\left\{\tilde{v}_i\right\}$), $i = 1\ldots \tilde{M}$ (resp. $\tilde{P}$), a basis of $\tilde{X}^h$ (resp. $\tilde{V}^h$) constructed from the non-zero traces of the w_i (resp. v_i) on Γ_3 . The fact that one uses P_1 finite elements enables us to assume the w_i to be positive. To each $\tilde{v}_i$ one can associate a unique element $R\tilde{v}_i = v_i$ on Γ_3 and equal to zero on each interior vertex of $\mathcal{T}^h$. This defines a linear mapping R from $\tilde{V}^h$ to V^h.

We now define a discrete version F^h and F^h_N of the mappings F and F_N by :

(22) $\quad < F^h(v), \tilde{v} > = a(v, R\tilde{v}) - L(R\tilde{v}) \qquad \forall \tilde{v} \in \tilde{V}^h, \quad \forall v \in V^h$,

(23) $\quad < F^h_N(v), \tilde{w} > = < F^h(v), \tilde{w}n > \qquad \forall \tilde{w} \in \tilde{X}^h, \quad \forall v \in V^h$.

Finally, let Π^h denote the restriction to Γ_3 of the interpolation operator associated with X^h. Note that Π^h is such that :

(24) $\quad \exists\, c(h): \quad \left|\Pi^h(|v^h|)\right|_{L^2(\Gamma_3)} \leqslant c(h)\, |v^h|_{L^2(\Gamma_3)} \qquad \forall v^h \in \tilde{V}^h$.

And once again because of the use of P_1 finite elements :

(25) $\quad \Pi^h(v)$ is positive for all positive continuous v on Γ_3 .

A discrete formulation of problem (P) in V^h is then :

(26) $$(P^h) \left[\begin{array}{l} \text{Find } u^h \in \mathbb{K}^h \text{ such that : } \forall v \in \mathbb{K}^h \\ a(u^h, v - u^h) - < \mu F^h_N(u^h), \Pi^h(|v_T| - |u^h_T|) > \geqslant L(v - u^h) . \end{array} \right.$$

Remark : In order to simplify the notation we suppose here that the friction coefficient μ is constant. The general case can be obtained by setting $\Pi^h\mu$ instead of μ.

3.2 <u>An equivalent fixed point problem</u> :

Let G be the set of positive linear mappings on $\tilde{X}^h$.
For all $g \in G$ the following problem :

(27) $$(P^h_g) \left[\begin{array}{l} \text{Find } u^h_g \in \mathbb{K}^h \text{ such that : } \forall v \in \mathbb{K}^h \\ a(u^h_g, v - u^h_g) + < g, \Pi^h(|v_T| - |u^h_{gT}|) > \geqslant L(v - u^h_g) . \end{array} \right.$$

has a unique solution. This is because the use of P_1 finite elements implies that the mapping $v \mapsto < g, \left|\Pi^h(|v_T|)\right| >$ is convex and therefore problem P^h_g is a classical minimization problem of a convex functional with quadratic growth.

Let us now define the mapping T by :

$$T(g) = -\mu\, F^h_N(u^h_g) \qquad \text{for } g \in G .$$

We shall now make sure that T takes its values in G, that is to say that T(g) is a positive linear mapping on $\tilde{X}^h$.

$$< T(g), \tilde{w}_i > = - < \mu\, F_N^h(u_g^h), \tilde{w}_i >$$

$$= \mu < F^h(u_g^h), - \tilde{w}_i n>$$

$$= \mu \left[a\,(u_g^h, R(-\tilde{w}_i n)) - L(R(-\tilde{w}_i n))\right] .$$

The $\tilde{w}_i$ are positive because of the use of P_1 finite elements, so that R $(-\tilde{w}_i n)$ is negative. By setting $v = u_g^h + R(-\tilde{w}_i n)$ (which is an element of $\mathbb{K}^h$) in inequality (27) one obtains :

$$a(u_g^h, R(-\tilde{w}_i n)) - L(R(-\tilde{w}_i n)) \geqslant - <g, \Pi^h(\left|(u_g^h + R(-\tilde{w}_i n))_T\right| - \left|u_{gT}^h\right|)> .$$

And finally as $(R(\tilde{w}_i n))_T = 0$,

$$< T(g), \tilde{w}_i > \geqslant 0 .$$

Therefore the existence of a fixed point of T is equivalent to the existence of a solution u^h of problem P^h.

3.3 <u>A constructive existence result in U^h</u>

We shall begin by establishing two lemmas concerning mapping T:

<u>Lemma 1</u> :There exists C(h), positive constant depending on h, such that , $\forall\ g_1, g_2 \in G$,

$$|T(g_2) - T(g_1)|_* \leqslant \mu\, C(h)\, |g_2 - g_1|_* ,$$

$$\text{where } |g|_* = \sup_{v^h \in \tilde{V}^h} \frac{|< g, v^h >|}{|v^h|_{H^{1/2}(\Gamma_3)}}$$

Proof :Taking g_1, $g_2 \in G$ and u_1, u_2 the corresponding solutions of $P^h_{g_1}$ (resp. $P^h_{g_2}$), by adding the two inequalities (27) one obtains :

$$a\,(u_2 - u_1, u_2 - u_1) \leqslant < g_2 - g_1, \Pi^h(\,|u_{2T}| - |u_{1T}|\,) > .$$

Because of the ellipticity of the elasticity matrix (13) one has :

$$|u_2 - u_1|_{L^2(\Gamma_3)} \leqslant \frac{1}{\alpha_0} |g_2 - g_1|_* \left|\Pi^h(|u_{2T}| - |u_{1T}|)\right|_{L^2(\Gamma_3)},$$

and by using (25) followed by (24) :

$$(28) \qquad |u_2 - u_1|^2_{L^2(\Gamma_3)} \leqslant \frac{c(h)}{\alpha_0} |g_2 - g_1|_* .$$

By the definition of F_N^h :

$$< F_N^h(u_1) - F_N^h(u_2), \tilde{w}_i > = a(u_1 - u_2, R\,\tilde{w}_i) ,$$

so that :

$$(29) \qquad \left|F_N^h(u_1) - F_N^h(u_2)\right|_* \leqslant C_1 \; |u_1 - u_2|_{H^{1/2}(\Gamma_3)} .$$

The lemma may now be easily deduced from (28) and (29) because the norms $H^{1/2}$ and L^2 are equivalent on the finite dimensional space $\tilde{V}^h$.

Lemma 2 : $\exists\, M > 0$ such that $|T(g)|_* \leqslant M \quad \forall\, g \in G$.

Proof : By setting v=0 in inequality (27) one obtains :

$$(30) \qquad \left|u_g^h\right|_{H^1} \leqslant C_1 ,$$

and because :

$$|T(g)|_* = \mu \sup_{v^h \in \tilde{V}^h} \frac{|< F_N^h(u_g^h), v^h >|}{|v^h|_{H^{1/2}(\Gamma_3)}}$$

$$= \mu \sup_{v^h \in \tilde{V}^h} \frac{\left|a(u_g^h, Rv^h n) - L(Rv^h n)\right|}{|v^h|_{H^{1/2}(\Gamma_3)}} ,$$

using (30) one concludes.

Theorem 1 : Problem P^h has a solution .

Proof : Let H be the intersection of G with the ball of centre 0 and radius M. H is a compact convex of the dual of $\tilde{V}^h$. Because of lemma 2, $T(H) \subset H$ and T is continuous by Lemma 1.

The existence of a solution u_g^h of P^h is deduced by applying Brower's theorem to the mapping T. Lemma 1 induces the following proposition.

Proposition : If $\mu < \dfrac{1}{C(h)}$, the mapping T is a contraction and there exists a unique fixed point g^* of T obtained as the limit of the following sequence :

$$\begin{cases} - \ g^0 \ \text{given} , \\ - \ g^{n+1} = T(g^n) . \end{cases}$$

Theorem 2 : If $\mu < \dfrac{1}{C(h)}$ problem P^h has a unique solution $u^h_{g^*}$.

4. Algorithm

We determine the fixed point of the application T introduced in section 3 .
For each given value of g, we solve problem P^h_g to get the solution u^h_g. Because of the symmetry of the bilinear form a(u,v), the variational inequality (27) is equivalent to the minimization of the functionnal $J^h(v)$ on the convex $\mathbb{K}^h$ with ,

$$J^h(v) = \frac{1}{2} a(v,v) - L(v) + < g, \Pi^h(|v_T|) > .$$

As before we introduce a mesh with P_1 triangles with linear interpolation for the displacements.

Denoting by I_N, I_T the set of the indices concerning the normal and tangential components of the contact displacements respectively, the convex $\mathbb{K}^h$ is written as :

$$\mathbb{K}^h = \prod_{i=1}^{2N} K_i , \qquad \text{with} \qquad K_i = \begin{cases} \mathbb{R}^- \ \text{if } i \in I_N , \\ \mathbb{R} \ \text{if } i \notin I_N . \end{cases}$$

N being the total number of nodes.

Note : this is due to the positiveness of the shape functions for P_1 elements. For higher order finite elements, $\mathbb{K}^h$ would be much more complicated.

We must find the vector $\bar{u}_g \in \mathbb{K}^h$ minimizing in $\mathbb{K}^h$ the functional:

$$J(\bar{v}) = \frac{1}{2} \bar{v}^T A \bar{v} - \bar{F}^T.\bar{v} + \bar{G}.\left|\bar{v}_T\right| ,$$

where - $\bar{v}$ is the vector of nodal displacements ,
- A is the finite element matrix of general term $a_{ij} = a(w_i, w_j)$ and where w_i are the shape functions ,
- $\bar{F}$ is the loading vector: $F_i = L(v_i)$,

- $\bar{G}$ is the sliding limit vector. We underline the essential role of the restriction mapping Π^h :

$$< g, \Pi^h\left(\left|\sum_i v_{Ti}\, w_i\right|\right) > = < g, \sum_i w_i\, |v_{Ti}| >$$

$$= \sum_i < g, w_i > |v_{Ti}| ,$$

$G_i = < g, w_i>$.

On the basis of the successive overrelaxation method(S.S.O.R.) we introduce the following specific algorithm which takes into account both the unilateral condition and the undifferentiable term. The minimization on one component v_i after the other leads to considering 3 cases :

If $i \notin I_N$ and $i \notin I_T$, we have the classical minimization of a quadratic functional on $\mathbb{R}$. The overrelaxation iteration is written :

$$u_i^{n+½} = \frac{1}{a_{ii}}\left(F_i - \sum_{j=1}^{i-1} a_{ij}\, u_j^{n+1} - \sum_{j=i+1}^{N} a_{ij}\, u_j^{n}\right),$$

$$u_i^{n+1} = (1-\omega)\, u_i^n + \omega\, u_i^{n+½} \quad (\omega \text{ is the overrelaxation coefficient}) .$$

We underline that the term $\bar{G}\left|\bar{v}_T\right|$ contains only components relative to indices belonging to I_T.

If $i \in I_N$, we have the minimization of the same functional on the convex $K_i = \mathbb{R}^-$,$u_i^{n+1}$is projected onto $\mathbb{R}^-$.

$$u_i^{n+1} = \text{Proj}_{R^-}\left((1-\omega)\, u_i^n + \omega\, u_i^{n+½}\right) .$$

If $i \in I_T$, we have the minimization on $\mathbb{R}$ of the functional including the absolute value which is simply treated by considering the two possibilities for u_i^{n+1} to be either positive or negative.

In Raous et al,[88], several acceleration procedures are given such as a diagonal process on the determination of g, a condensation of the problem to the contact variables alone, and a special matrix storage. So as not to alterate the study of the convergence of the fixed point method presented in the next paragraph, we will not use the diagonal process wich consists in partial resolution for the first determinations of g.

We use an optimum relaxation coefficient determined through a dichotomic research.

5. Where results of section 3 are tested

To begin with we note that, applied to models connected to physical problems, a good behaviour of the algorithm and of its convergence has been observed.

We will focus here on how the convergence of the fixed point iteration is influenced by the value of the friction coefficient (even for unrealistic values), by mesh discretization, and by the choice of the initial condition g^0 .

The discussion is not exhaustive because it is limited to two examples and to relatively few meshes; however it suggests a few comments on the behaviour of the algorithm with regard to the results of section 3.

5.1 Description of the examples

Example A : contact of a long bar with a plane surface (plane strain hypothesis).

A section of the bar is represented on Fig. 1 . A force density $f = -5$ daN/mm^2 is applied on the side GE, and a force density $F = 10$ daN/mm^2 is applied on AG. On the side AD, unilateral conditions with Coulomb friction are assumed. On the side ED, boundary conditions are imposed to assume the symmetry ($u_x = 0$) and to fix the point D ($u_x = u_y = 0$). Plane strain elasticity is assumed with a Young modulus E equal to 13 000 daN/mm^2 and a Poisson coefficient ν equal to 0.2.

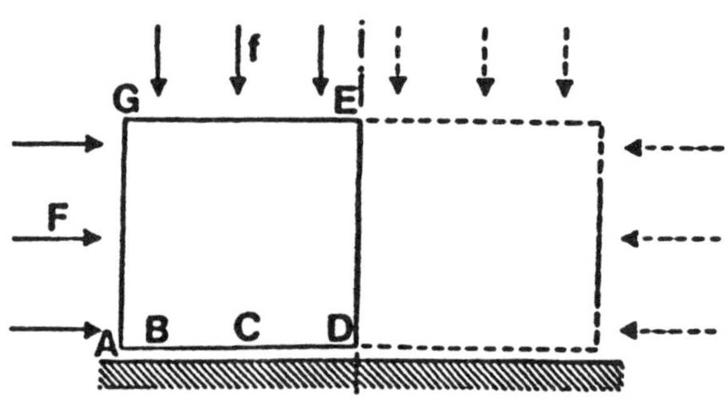

Fig. 1 : geometry of example A .

Details of the solution for different values of μ are given in Raous et al [88].

The test will be done on five different meshes . On Fig. 2, the mesh A1 has a local refinement on the contact boundary :N= 230 and NC= 33. N denotes the total number of nodes and NC the number of contact nodes.On Figs. 3 to 6, four uniform meshes are given denoted by A2, A3, A4, A5.

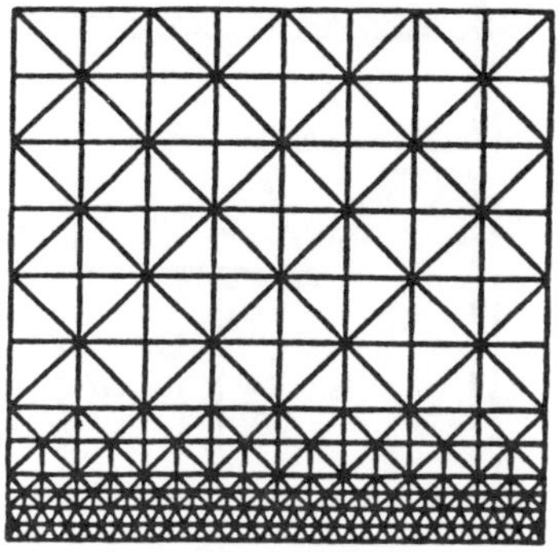

Fig. 2 : mesh A1 of example A ,
N = 230 , NC = 33 .

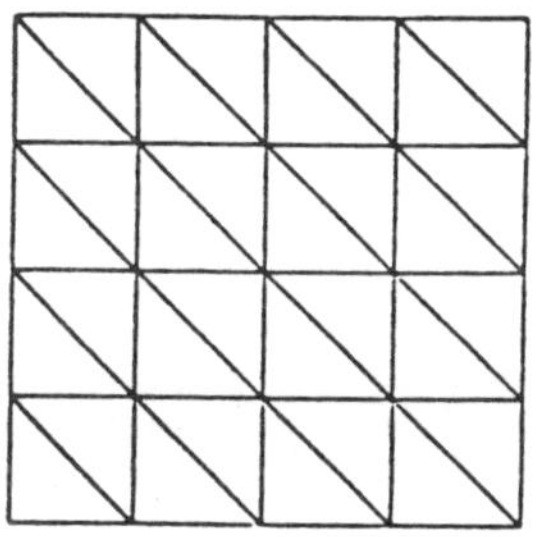

Fig. 3 : mesh A2 of example A ,
N = 25 , NC = 5 .

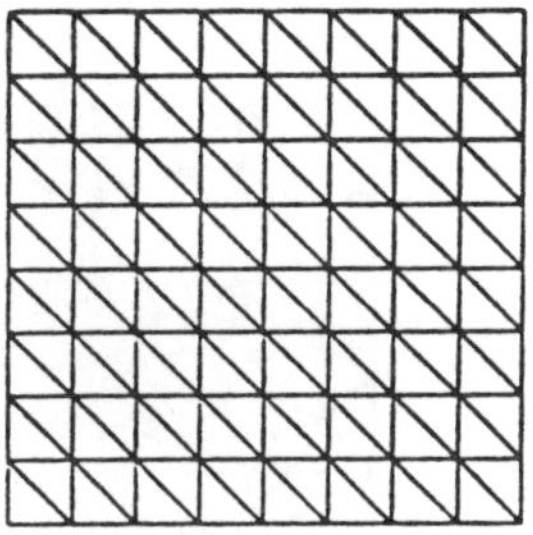

Fig. 4 : mesh A3 of example A,
N = 81 , NC = 9 .

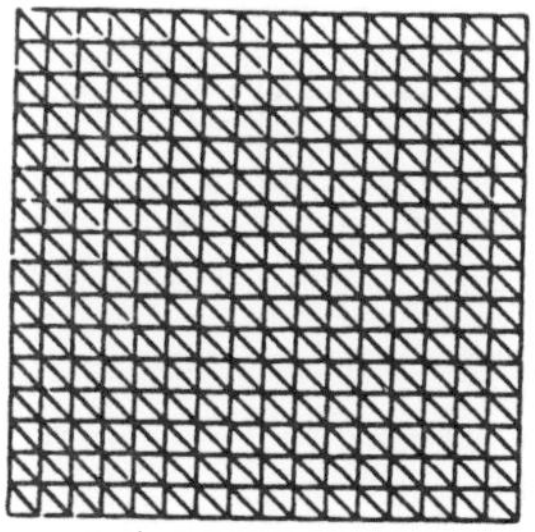

Fig. 5 : mesh A4 of example A ,
N = 289 , NC = 17 .

Fig. 6 : mesh A5 of example A,
N = 1089 , NC = 33 .

Example B : a dovetail assembling .

The geometry is given in Fig. 7. This is a two body contact problem with an oblique contact zone. We show in Raous et al ,[88], how to generalize the formulation and the algorithm for such a problem. In the same paper, the complete solution is given. Plane stress hypothesis is assumed. The meshes given in Fig. 8 are used (they have respectively 19 and 9 couples of contact nodes).

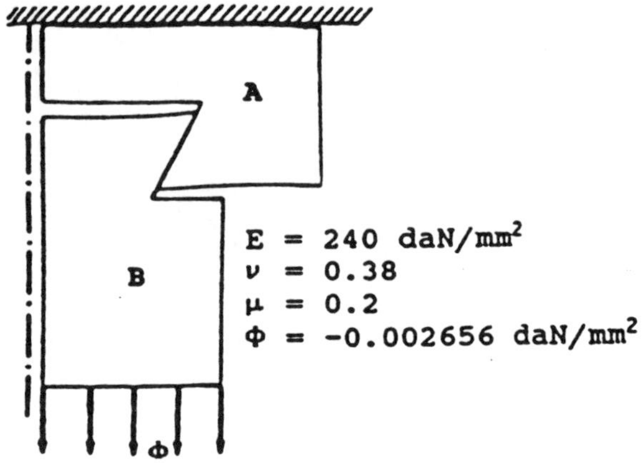

Fig. 7 : example B .
Dovetail assembling (displacements amplified twenty times).

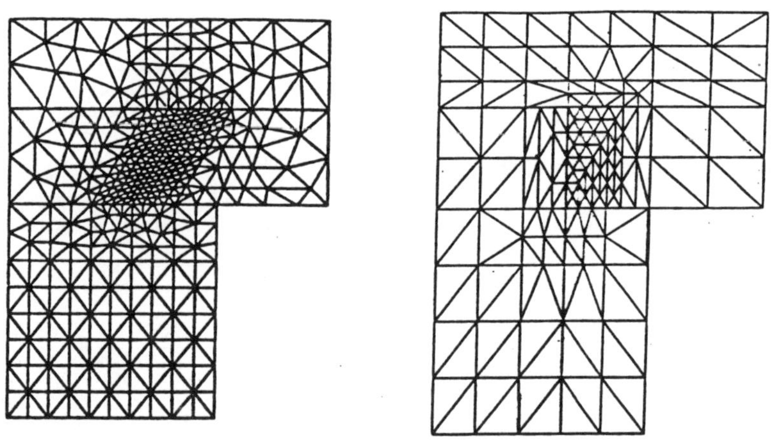

Fig. 8 :meshes B1 (N=384 NC=38) and B2 (N=148 NC=18) of example B .

5.2 <u>The Influence of the magnitude of μ</u>

In the next two tables we give the number of iterations on g needed for the fixed point process to converge for different values of μ and for different meshes. Some values of μ are very large and have no mechanical sense : the point is to check the apparent restriction of theorem 2 on the magnitude of μ.

Table I concerns example A and table II concerns example B.

The indication "no convergence" means that the tests of convergence were not satisfied after a given number of iterations depending on the size of the problem.

The test of convergence on g is :

$$\frac{|g^{k+1} - g^k|_\infty}{|g^k|_\infty} < 10^{-3} \quad , \qquad \text{with } |g|_\infty = \sup_{i \in I_T} |g_i| \; .$$

The test of convergence for the relaxation procedure is :

$$\frac{\|u^{k+1} - u^k\|_1}{\|u^k\|_1} < .5\ 10^{-5} \quad , \quad \text{with} \quad \|u^k\|_1 = \sum_{i=1}^{2N} |u_i^k| .$$

mesh / value of μ	A1	A2	A3	A4	A5
0.01	4	4	4	4	4
0.1	6	5	5	6	6
0.2	7	6	6	6	7
0.4	7	7	7	8	8
0.6	8	8	8	8	9
1	10	9	9	9	10
2	14	11	11	12	15
5	no conv	5	no conv	no conv	no conv
50	8	5	5	no conv	no conv
100	8	5	5	no conv	no conv

Table I: number of fixed point iterations for different meshes of example A.

mesh / value of μ	B1	B2
0.01	4	4
0.1	6	7
0.2	9	9
0.4	21	22
0.6	no conv	no conv
1	no conv	no conv
2	no conv	no conv
5	14	no conv
10	22	no conv
50	6	no conv
100	6	no conv

Table II: number of fixed point iterations for different meshes of example B.

Comments :

* In the range of physical admissible values of the friction coefficient, the solution is obtained in less than 10 iterations on g and the influence of μ is weak for this range of values. When the problem is close to a frictionless one the number of iterations to get convergence is small (4 iterations), in other cases it is about twice as much (between 6 and 10 iterations).

* For large values of μ (between 1 and 100), convergence is not obtained for special ranges which are not especially the largest values of μ. By checking the values of g in these cases, we observe that g oscillates between two neighbouring values: the difference resides in one or two nodes located at the boundary between the separated zone and the sliding one.Theses nodes are oscillating between contact with small contact force (u_N=0, F_N small) and a small separation (u_N small,F_N=0). The first state introduces a small tangential force whereas the second one gives a zero tangential force. The difference ($g^{k+1} - g^k$) does not succeed in passing the convergence test(10^{-3}), and the algorithm does not stop. This behaviour seems to be associated to the difficulty to compute the limit case of contact without transmission of normal forces ($u_N = 0$, $F_N = 0$). This has been observed also on an example presented in Raous et al [88] concerning a metal forming process with a small friction coefficient ($\mu = 0.2$).

* The mesh seems to have some influence because cases with large fricion coefficients (μ = 100 for example) converge for some grids and not for others. The limit between the separated zone and the clamped zone (for μ = 100 there is no sliding zone) introduces a singularity of the same kind as that of a crack in an elastic medium and that we are using values of g computed in this part.

We have to underline that the presence of a sliding zone introduces a tangential force in the neighbourhood of the clamped part and gives a regularization of the previous singularity. The solution is smoother for small friction coefficient and the convergence is the same for the different meshes.

5.3 Influence of the choice of the initial condition g^0

As presented in table III, the choice of the initial condition g^0 seems to have no influence on the number of iterations on g. TableIII concerns example A with μ = 1 treated on the mesh A1, for which a good convergence has been observed.

initial condition g^o	number of iterations on g	Total number of relaxation iterations
(0,0,... ,0)	10	207
(1,1,... ,1)	10	205
(10,10,... ,10)	9	172
(100,100,... ,100)	9	172
(solution computed for μ = 0.6)	9	172
(small perturbation of the right solution)	9	171

Table III : influence of the initial condition g^0 on the convergence for example A with μ = 1 treated on the mesh A1 (N = 230, NC = 33).

5-4 Influence of the mesh on the convergence and on the solution

Fig. 9 gives a representation of the contact boundary of the solution of example A for the different meshes. A good coherence of the solutions is observed. One observes numerically the convergence of the solution as h decreases, for the contact aera.

mesh A2	0 x x I I
mesh A3	0 x x x x I I I I
mesh A4	0 0 x x x x x x x I I I I I I I I
mesh A5	0 0 0 x x x x x x x x x x x x x x I I I I I I I I I I I I I I I I
mesh A1	0 0 0 x x x x x x x x x x x x x x I I I I I I I I I I I I I I I I

Fig. 9 : representation of the contact boundary for example A ($\mu = 1$) for the different meshes : 0 separate node
x sliding node
I stick node .

6. Conclusions

We submit in this article a dicretization by classical finite elements P1 of the unilateral contact problem with Coulomb friction. Existence, as well as uniquenesss when the value of the friction coefficent μ is small, has been obtained. A fixed point theorem is used to establish unicity for μ small, therefore convergence of the successive approximation algorithm follows. The use of a classical finite element method implies that the implementation is simple (at each iteration a problem of minimization of a convex functional is solved) and therefore allows for many numerical tests. One observes in particular that for realistic values of the friction coefficent convergence is obtained in less than ten iterations. For these same values of μ the discrete solution behaves very well with regard to the mesh refinement, which seems to indicate that the solution of the discrete problem has a limit when the mesh step tends to zero .

REFERENCES

Ciarlet P.: The finite element method for elliptic problems, North Holland, 1978.

Cocu M.: Existence of solutions of Signorini problems with friction, 22, n°5, 567-575, 1984.

Duvaut G.: Equilibre d'un solide élastique avec contact unilateral et frottement de Coulomb, CRAS, Sèrie A, 290, 263-265, 1980.

Duvaut G.,Lions J.L.: Les inéquations en mécanique et en physique, Dunod, Paris, 1972.

Haslinger J.: Approximation of the Signorini problem with friction, obeying the Coulomb law, Math.Meth. in Appl. Sci. 5, 422-437, 1983.
Jarusek J.:Contact problem with a bounded friction. Coercive case, Czech. Math. J., 33, 2, 254-278, 1983.
Jeannin L.: Etude mathématique et numérique d'un problème de contact unilatéral avec frottement en élasticité et élastoplasticité, thèse de docteur ingénieur, école centrale de Lyon, 1985.
Klarbring A., Mikelic A. and Shillor M.: Frictional contact problems with normal compliance, Int. J. Engng. Sci., Vol 26, n°8, 811-832, 1988.
Necas J., Jarusek J., Haslinger J.: On the solution of the variational inequality to the Signorini problem with small friction, Bollettino U.M.I., 5, 17 B, 796-811, 1980.
Oden J.T. and Martins J.A.C.: Model of computational methods for dynamic friction phenomena, Comp. Meth. Appl. Mech. Eng., Vol 52, 527-634, 1985.
Raous M., Chabrand P., Lebon F.: Numerical methods for frictional contact problems and applications, Journal of theoretical and applied mechanics, Special issue, supplement n°1 to vol 7, 111-128, 1988.
Temam R.: Problèmes mathématiques en plasticité, Gauthiers-Villars, Paris, 1983.

* : L.G.M.C.-U.S.T.L. - 34095 Montpellier cedex 2- France.
** : L.M.A.-C.N.R.S. - 13402 Marseille cedex 9 -France.

RECENT RESEARCH ON ELASTIC-PLASTIC OSCILLATORS

David R. Owen

Department of Mathematics, Carnegie Mellon University,
Pittsburgh, PA, U.S.A.

Introduction

In the past twenty years an increasing amount of attention has been devoted to research on elastic-plastic oscillators. These mechanical systems are useful model systems for studies of the complicated behavior of three-dimensional bodies, because they are governed by differential equations that mathematically are of independent interest and that have solutions displaying a rich variety of physically interesting characteristics. Knowledge of these systems provides a point of reference for studying the partial differential equations that arise in more detailed models of elastic-plastic behavior, and solutions of the ordinary differential equations describing coupled systems of elastic plastic oscillators are candidates for lumped approximations of solutions of the partial differential equations. Moreover, elastic-plastic oscillators are useful tools in the design of structures that must withstand large ground motions due to earthquakes, because these oscillators are relatively easy to simulate numerically, and their motions display some of the features of structures that are severely deformed. References [1]-[12] provide a selection of examples of recent research on elastic-plastic oscillators.

In this paper I wish to outline a few recent results on elastic-plastic oscillators that have been obtained at Carnegie Mellon as part of an ongoing study. A common theme underlying this study is the explicit identification of appropriate mathematical settings for describing each mechanical system under consideration. Following this theme permits us not only to state our results with precision, but also to point out interesting phenomena, such as the accumulation of transition times between elastic and plastic intervals of behavior backward in time in the case of an elastic-perfectly plastic oscillator with integrable forcing function. This possibility complicates and renders more interesting and challenging the task of establishing theorems on existence, uniqueness, and continuous dependence of solutions upon initial data for such an oscillator. The results that I shall discuss here are organized around the specific topics:

1) critical damping and mechanical self-annealing in unforced motions of a bilinear, elastic-plastic oscillator;
2) weakly decaying energy separation and uniqueness of motions of an elastic-plastic oscillator with work-hardening;
3) resonance in elastic-perfectly plastic oscillators.

Critical damping and mechanical self-annealing

In references 7 and 11, John Thomas and I have studied the behavior for large times of an elastic-perfectly plastic oscillator subject to a linear restoring force. This mechanical system also has been studied, particularly in the engineering literature, in an equivalent formulation as an unforced bilinear elastic-plastic oscillator. From either point of view, one can visualize a rigid mass that is supported both by a linear spring and an elastic-perfectly plastic rod, or filament, and that undergoes one-dimensional motions. The total displacement $u = u_e + u_p$ and the velocity $v = \dot{u}$ of the mass as functions of time are governed by the following system:

$$|u_e(t)| \leq \alpha, \tag{1}$$

$$m\dot{v}(t) = -k(u_e(t)+u_p(t)) - \mu u_e(t), \tag{2}$$

$$\dot{u}_e(t) = \begin{cases} 0 & \text{if } |u_e(t)| = \alpha \text{ and } \\ & \quad u_e(t)v(t) \geq 0 \\ v(t) & \text{otherwise,} \end{cases} \tag{3}$$

$$\dot{u}_p(t) = \begin{cases} v(t) & \text{if } |u_e(t)|=\alpha \text{ and } \\ & \quad u_e(t)v(t) \geq 0 \\ 0 & \text{otherwise.} \end{cases} \tag{4}$$

Here, the positive number α is called the <u>yield displacement</u>, u_e is called the <u>elastic displacement</u>, and u_p is called the <u>plastic displacement</u>. The positive numbers k and μ are the stiffness of the linear spring and of the elastic-plastic filament, respectively. Because $v=\dot{u}$, relations (1), (3) and (4) can be used in priciple to determine u_e and u_p on each interval of time on which the total displacement $u=u_e+u_p$ is prescribed; likewise, the total force $F = (k+\mu)u_e+ku_p$ on the mass is determined by u on such intervals. The parametric curve $t \mapsto (u(t),F(t))$ so obtained has slope dF/du given by:

$$\frac{dF}{du} = \begin{cases} k & \text{if } |u_e| = \alpha \\ & \quad \text{and } u_e\dot{u} \geq 0 \\ k+\mu & \text{otherwise,} \end{cases} \tag{5}$$

so that the force-displacement curve is composed of line segments of slopes $k+\mu$ and k, which has led to the terminology <u>bilinear elastic-plastic oscillator</u>, or <u>bilinear hysteretic oscillator</u>.

Thomas and I show ([11], Theorem 3.1) that, for every initial state $(v^\circ,u_e^\circ,u_p^\circ)$ in the strip

$$\mathcal{S} = \{(v,u_e,u_p) \in \mathbb{R}^3 \mid |u_e| \leq \alpha\}, \tag{6}$$

there is exactly one continuous, locally piecewise continuously differentiable function $(v,u_e,u_p) : [0,\infty) \to \mathcal{S}$ that satisfies (1)-(4) at all points of differentiability and the initial condition

$$(v(0),u_e(0),u_p(0) = (v^\circ,u_e^\circ,u_p^\circ); \tag{7}$$

moreover, this function can be obtained by patching together "elastic segments", i.e., solutions of the system (1), (2), $(3)_2$, and $(4)_2$, and "plastic segments", i.e., solutions of (1), (2), $(3)_1$, and $(4)_1$. In particular, this result implies that for such an oscillator transition times between elastic and plastic segments only can accumulate at $+\infty$, and that each solution trajectory for the initial value problem (1)-(4), (7) <u>locally</u> is a planar curve contained in a plane $u_e = \pm\alpha$ or in a plane u_p = constant in v, u_e, u_p space. Our main interest in studying unforced motions of the bilinear elastic-plastic oscillator was to gain an understanding of the behavior of solution trajectories as $t \to +\infty$ and, more specifically, to understand the behavior of $u_p(t)$, the plastic displacement at time t, as $t \to +\infty$. We were able to obtain a complete description of these behaviors, and the rest of this section is devoted to a summary of our most striking results, described in detail in the article [11], Sections 4-6.

In order to simplify the discussion, I consider only initial states of the form $(\overset{\circ}{v}, \overset{\circ}{u}_e, 0) \in \mathcal{S}$, so that there is no initial plastic displacement of the oscillator. A first result on trajectories with such initial states ([11], Theorem 5.1) is the assertion: <u>every trajectory approaches an ellipse in a plane of the form</u> $u_p = u_\infty$; <u>if the ellipse is not reached in finite time, then</u> $u_\infty = 0$ <u>and the ellipse is given by</u>

$$\frac{1}{2}mv^2 + \frac{1}{2}(\mu+k)u_e^2 = \frac{1}{2}(\mu+k)\,\alpha^2. \tag{8}$$

In other words, if the plastic displacement u_p does not attain its limiting value u_∞ in finite time, then u_∞ is zero and the trajectory approaches the ellipse (8) as $t \to +\infty$. Actually, Thomas and I were able to show that the ellipse (8) is approached in an oscillatory manner, in that the sign of u_p on elastic segments of the trajectory must change from one elastic segment to the next. It is natural to refer to this decaying, oscillatory behavior of u_p as an <u>underdamped oscillation</u>, because this term is used to describe a similar decaying, oscillatory behavior of the total displacement of a linearly damped harmonic oscillator. When underdamped oscillations do not occur for a solution of (1)-(4), (7), then u_p necessarily reaches its final value u_∞ in finite time, and, of course, u_p does not change sign thereafter, so that it is natural to use the term overdamped oscillation to describe this case.

The main result on trajectories with initial states of the form $(v^\circ, u_e^\circ, 0)$ shows that the stiffness ratio

$$\kappa = k/(k+\mu) \tag{9}$$

of the bilinear elastic-plastic oscillator determines whether underdamped or overdamped oscillations occur. Specifically, Thomas and I proved (see [11], Section 6): If $0 < \kappa < \frac{1}{2}$, i.e., if $k < \mu$, then only overdamped oscillations can occur, so that every trajectory enters a plane $u_p = u_\infty$ in finite time. If $\frac{1}{2} \le \kappa < 1$, i.e., if $\mu \le k$, then only underdamped oscillations can occur, so that every trajectory either starts and remains in the plane $u_p = 0$ for all times or leaves the plane $u_p = 0$ at some time and approaches the ellipse (8) in the plane $u_p = 0$ with oscillatory values of u_p as $t \to \infty$. Another way of regarding the underdamped case is embodied in the assertion: plastic displacements, no matter how large, that arise during motions of an oscillator with $\frac{1}{2} \le \kappa < 1$, automatically disappear in the limit of large times. Thomas and I call this behavior self-annealing (or mechanical self-annealing) in motions of the oscillator. (The adjective "mechanical" helps to emphasize that the removal of plastic displacements occurs in the absence of thermal devices.) Moreover, we say that $\frac{1}{2}$ is a critical stiffness ratio for the oscillator, because this ratio separates the overdamped and underdamped cases. An additional result that we obtained in [11] for the overdamped case (in which u_∞ is not necessarily zero), is the following bound on u_∞ that is both sharp and independent of initial state:

$$|u_\infty| \leq \alpha \left[\frac{1}{\kappa} - 2\sqrt{\frac{1}{\kappa} - 1} \right]. \tag{10}$$

Therefore, whatever the case may be regarding underdamped or overdamped oscillations, sharp bounds are obtained on the ultimate plastic displacement that are independent of the initial state $(v^{\circ}, u_e^{\circ}, 0) \in \mathcal{S}$.

Thomas and I use geometrical and analytical arguments to establish the foregoing results on asymptotic behavior of trajectories of the system (1)-(4). These arguments exploit the piecewise planar character of trajectories and are phrased to a large extent in the language used to study autonomous systems of ordinary differential equations. Jacobo Bielak has pointed out to us how some of our results can be obtained by means of graphical arguments using the bilinear character of the force-displacement curves, and he has done numerical calculations, both for the oscillators in question as well as for these oscillators with additional linearly viscous damping, that illustrate our results. These ideas and calculations are described in the preprint [12] from the point of view of earthquake engineering, in which the oscillator in its initial state is viewed as a simplified model for a structure that is loaded by ground motion of short duration.

Weakly decaying energy separation and uniqueness of motions

It is instructive to consider the bilinear elastic-plastic oscillators described by (1)-(4) as elastic-perfectly plastic oscillators subject to linear conservative loading. In fact, if we put $\sigma = \mu u_e$ and $u = u_e + u_p$ in the system (1)-(4), we obtain the system

$$|\sigma(t)| \leq \mu\alpha, \tag{11}$$

$$m\dot{v}(t) = -ku(t) - \sigma(t), \tag{12}$$

$$\dot{\sigma}(t) = \begin{cases} 0 & \text{if } |\sigma(t)| = \mu\alpha \text{ and } \\ & \sigma(t)v(t) \geq 0 \\ \mu v(t) & \text{otherwise,} \end{cases} \tag{13}$$

$$\dot{u}_p(t) = \begin{cases} v(t) & \text{if } |\sigma(t)| = \mu\alpha \text{ and } \\ & \quad \sigma(t)v(t) \geq 0 \\ 0 & \text{otherwise,} \end{cases} \tag{14}$$

In this system, σ represents the force on the mass exerted by an elastic-perfectly plastic filament with yield force $\sigma_y = \mu\alpha$, and the term $-ku$ in (12) represents a conservative, linear loading on the mass (in addition to the force σ on the mass due to the filament). This change in point of view leads naturally to various ways in which more general oscillators can be introduced: one can replace the conservative force $-ku(t)$ by an external force of the form $f(t,u(t))$ and/or one can allow the yield stress $\mu\alpha$ and the stress and plastic-displacement rates to have more general forms than those set forth in (11), (13), and (14). In the rest of this paper I consider an external force acting on the mass of the form $f(t)$ and the following more general forms of (11), (13) and (14):

$$|\sigma(t)| \leq \sqrt{2H(w(t))}, \tag{15}$$

$$\dot{\sigma}(t) = \begin{cases} \dfrac{\mu H'(w(t))}{\mu + H'(w(t))} v(t) & \text{if } |\sigma(t)| = \sqrt{2H(w(t))} \\ & \text{and } \sigma(t)v(t) \geq 0 \\ \mu v(t) & \text{otherwise,} \end{cases} \tag{16}$$

$$\dot{w}(t) = \begin{cases} \dfrac{\mu\sigma(t)}{\mu + H'(w(t))} v(t) & \text{if } |\sigma(t)| = \sqrt{2H(w(t))} \\ & \text{and } \sigma(t)v(t) \geq 0 \\ 0 & \text{otherwise.} \end{cases} \tag{17}$$

The smooth function H describes how the yield stress can vary with the state of the oscillator; the quantity w, called the plastic work, derives its name from the relation

$$\dot{w}(t) = (v(t) - \frac{\dot{\sigma}(t)}{\mu})\sigma(t)$$
$$= (\dot{u}(t) - \dot{u}_e(t))\sigma(t) = \sigma(t)\dot{u}_p(t), \tag{18}$$

and (15)-(17) embody the requirement that the bound on stresses (15) depends upon the amount of plastic work that the oscillator has experienced. In fact, if $(18)_1$ is taken as the definition of $\dot{w}$, then $(16)_1$ follows from $(18)_1$ and the relation (15), with equality instead of inequality. When H is a constant function, in view of (18), the relations (15), (16), (17) reduce to (11), (13), and (14). The elastic-plastic oscillator described by (15)-(17) together with the momentum equation

$$m\dot{v}(t) = f(t)-\sigma(t) \tag{19}$$

is called an <u>elastic-plastic oscillator with (isotropic) work-hardening</u>; when H is a constant, it is called an <u>elastic-perfectly plastic oscillator</u>.

In this section, I describe some complications that arise in proving uniqueness theorems for elastic-plastic oscillators with work hardening when the forcing function f in (19) is allowed to be an integrable function, with no other restrictions imposed on the rate at which the values of f can oscillate. The complications arise for two reasons: first of all, the method of patching together elastic and plastic segments to form a solution trajectory, as described briefly in the previous section, no longer applies to the case of arbitrary, integrable forcing functions f, so that one must allow for the possibility that transition times separating elastic and plastic segments accumulate backwards in time; second of all, the variation of yield stress with state permits the oscillator to store energy during plastic deformations, a situation that is ruled out for the elastic-perfectly plastic oscillator, so that the dissipation due to plastic deformation is weakened by the presence of work hardening. A qualitative theory of the system (15)-(17), (19) when H is a constant and f is locally integrable was given by Buhite and me in the article [5]; in particular, the proof of uniqueness of classical solutions of the initial value problem for an elastic-perfectly plastic oscillator was facilitated by the inequality

$$\dot{e}(t) = \left(\frac{1}{2}m(v-\bar{v})^2 + \frac{1}{2\mu}(\sigma-\bar{\sigma})^2\right)^{\bullet}(t) \leq 0, \tag{20}$$

according to which the <u>energy separation</u> e of two solutions (v,σ,w) and $(\bar{v},\bar{\sigma},\bar{w})$ is a non-increasing function of time. Indeed, when H is a constant function, (17) uncouples from (15), (16), and (19), and relations (20) and (18) then immediately imply uniqueness of solutions of the initial value problem for elastic-perfectly plastic oscillators.

For an elastic-plastic oscillator with work-hardening, H is not a constant, and inequality (20) is replaced by the inequality

$$\dot{e}(t) \leq -(\sigma(t)-\bar{\sigma}(t))(\dot{u}_p(t)-\dot{\bar{u}}_p(t)) \tag{21}$$

in which the right-hand side need not be of one sign. In fact, the system (15)-(17), (19) does not satisfy a classical condition such as monotonicity or one-sided Lipschitz continuity that would yield uniqueness of solutions. Nevertheless, I was able to prove in the paper [9] that, although e need not decrease with time, as would be the case if (20) were to hold, the energy separation e has the following <u>property of weak decay</u>: if $u_p(0)=\bar{u}_p(0)$, then for every $t\geq 0$,

$$e(t) \leq e(0). \tag{22}$$

Thus, although e can increase over certain intervals, any such increase must be compensated for by a prior decrease in such a way that (22) remains valid for all times. The condition $u_p(0)=\bar{u}_p(0)$ arises in connection with (22), because the proof of (22) given in [9] uses an auxillary system of equations in which u_p plays a role similar to that of w in (15)-(17), (19). Actually, the condition $u_p(0)=\bar{u}_p(0)$ can be met and relation (18) remains valid if we define u_p in terms of v and σ,

$$u_p(t) = \int_0^t (v(\tau) - \frac{\dot{\sigma}(\tau)}{\mu}) d\tau, \tag{23}$$

whenever v and σ are the initial entries in a solution (v,σ,w) of the system (15-17), (19). Once one has proven (22), it follows immediately that if (v,σ,w) and $(\bar{v},\bar{\sigma},\bar{w})$ are two solutions of (15)-(17), (19) with the same initial data, then $e(0) = 0 = e(t)$ for all $t \geq 0$, so that $v=\bar{v}$ and $\sigma=\bar{\sigma}$. Therefore, by (18), $w=\bar{w}$, and uniqueness of solutions is established.

Keming Wang and I recently have used the ideas in the above argument to obtain the following uniqueness theorem for systems of ordinary differential equations.: Let a positive number T, two positive integers m, n with $m<n$, and $x^0 \in \mathbb{R}^n$ be given, and assume that $f=(f_1, f_2, \ldots, f_n): \mathbb{R}^n \times \mathbb{R} \to \mathbb{R}^n$ satisfies the condition: there is an integrable function $L : [0,T) \to \mathbb{R}^+$ and, for each $j=m+1, \ldots n$, not only is f_j non-negative, but also there is a concave, strictly increasing function $G_j: \mathbb{R} \to \mathbb{R}$ such that

$$(Px-P\bar{x}) \cdot (Pf(x,t)-Pf(\bar{x},t)) + \sum_{j=m+1}^{n} (G_j(x_j)-G_j(\bar{x}_j))(f_j(x,t)-f_j(\bar{x},t)) \leq L(t) \|Px-P\bar{x}\|^2, \tag{24}$$

whenever (x,t), $(\bar{x},t)$ lie in a given open subset of $\mathbb{R}^n \times \mathbb{R}$ containing $\{x^0\} \times [0,T)$. (Here, $Px=(x_1,\ldots,x_m)$ and $Pf(x,t) = (f_1(x,t),\ldots,f_m(x,t))$, so that P is a projection from $\mathbb{R}^n$ onto $\mathbb{R}^m$.) It follows that the initial value problem

$$\left.\begin{aligned} \dot{x}(t) &= f(x(t),t), \quad \text{a.e. } t \in [0,T) \\ x(0) &= x^0, \end{aligned}\right\} \tag{25}$$

has at most one absolutely continuous solution.

A simpler example than the elastic-plastic oscillator to which this theorem applies is the damped non-linear oscillator

$$\ddot{x}(t) = -x^{1/3}(t) - \dot{x}(t), \tag{26}$$

or, equivalently, the system

$$\begin{aligned} \dot{x}_1 &= -x_2^{1/3} - x_1 \\ \dot{x}_2 &= \qquad\quad +x_1 . \end{aligned} \tag{27}$$

When $x_1(0)>0$ and $x_2(0)=0$, uniqueness of solutions does not follow from classical uniqueness results based on (one-sided) Lipschitz continuity or on monotonicity of the right-hand sides, but the above uniqueness theorem, with $n=2$, $m=1$, $L=0$, and $G_2(x_2) = x_2^{1/3}$, does yield uniqueness of solutions for such data. A proof of that theorem as well as other applications will appear elsewhere. That proof rests on an inequality that permits one to remove the sum on the left-hand side of (24) after integration of both sides along solution curves; the resulting integrated form of (24) then yields uniqueness of the projection Px of a solution x of (25), and (24) again is invoked to obtain uniqueness of x. I recently learned of an article [13] by Gröger, Necas, and Trávnicek (based on earlier work of Moreau and of Halphen and Nguyen) in which the theory of monotone operators was adapted and successfully applied to dynamical problems in plasticity. In spite of the fact that the system (15)-(17), (19) does not, as a whole, define a monotone operator, the arguments in that article can be applied to this system to yield the weak decay of e and, hence uniqueness of solutions. These monotonicity arguments rest on the appearance in part of the system (15-17), (19) of a monotone operator, whereas the arguments used by Wang and me rely on the disappearance from the left-hand side of (24) of terms in $x_{m+1},\ldots,x_n$ through cancellation with other terms whose integrals obey a sign condition.

Resonance in elastic-perfectly plastic oscillators

In this section I consider the system (15)-(17), (19) in the special case where no hardening is present, i.e., when H is a constant:

$$|\sigma(t)| \leq \sigma_y \tag{28}$$

$$\dot{\sigma}(t) = \begin{cases} 0 & \text{if } |\sigma(t)|=\sigma_y \\ & \text{and } \sigma(t)v(t)\geq 0 \\ \mu v(t) & \text{otherwise,} \end{cases} \tag{29}$$

$$\dot{w}(t) = \begin{cases} \sigma(t)v(t) & \text{if } |\sigma(t)|=\sigma_y \\ & \text{and } \sigma(t)v(t)\geq 0 \\ 0 & \text{otherwise,} \end{cases} \tag{30}$$

$$m\dot{v}(t) = f(t) - \sigma(t). \tag{31}$$

Here, the positive number $\sigma_y = \sqrt{2H}$ is the <u>yield stress</u>, and, as noted in the previous section, the equation (30) uncouples from the system (28), (29), (31). Buhite and I ([5]) studied (28), (29), (31) in some detail. Among our results was a characterization of the <u>periodic</u> forcing functions f that admit periodic solutions of (28), (29), (31), i.e., solutions (v,σ) with the same period as f. Note that periodicity of v and σ does not imply periodicity of the displacements u, and I wish to discuss briefly some ongoing research by Louis Blair and me on the existence and stability of periodic solutions (v,σ) of (28), (29), (31) for which u also is periodic. This condition on u can be stated as a condition on $v=\dot{u}$:

$$\int_0^T v(t)dt = 0, \tag{32}$$

where T is the period of f, v, and σ. Of particular interest is the case where the forcing function f is given by

$$f(t) = f_0 \sin \left(\sqrt{\frac{\mu}{m}}\, t\right). \tag{33}$$

This forcing function causes "elastic resonance", in the sense that all solutions of $(29)_2$ and (31) are unbounded as $t \to +\infty$; moreover, the displacements associated with each such solution are unbounded. Nevertheless, Theorems 7.1 and 8.1 in [5] show that (28), (29), (31) does have a unique periodic solution (v_r,σ_r) when f is given by (33), so that the dissipation afforded by plastic deformations is strong enough to keep the velocity bounded in spite of elastic resonance. Although Theorem 6.1 in [5] does give insight in some cases into the behavior of displacements, that result yields no information on the displacement u_r for the solution (v_r,σ_r). Louis Blair recently has been able to show that if $|f_0|$ is sufficiently small the velocity v_r does indeed obey (32), so that the displacements u_r also are periodic. His argument uses an explicit guess as to the form of v_r when $|f_0|$ is small, and he then shows that there is a solution trajectory $t \mapsto (v(t),\sigma(t))$ for (28), (29), (31) that satisfies

$$(v(\pi/\sqrt{\mu/m}),\ \sigma(\pi/\sqrt{\mu/m})) = -(v(0),\sigma(0)), \quad (34)$$

from which it follows that $(v,\sigma) = (v_r,\sigma_r)$ and (32) is satisfied. We believe that this conclusion is valid for a broader range of values of $|f_0|$ than those treated in Blair's proof, and we are studying this question at present.

Blair's result on the periodicity of the displacements for v_r shows that the dissipation afforded by an elastic-perfectly plastic filament is strong enough to overcome the effects of elastic resonance. A second result of Blair concerns the displacements of non-periodic (v,σ) that are solutions of (28), (29), (31), with f as in (33). Theorem 8.1 of [5] tells us that $(v(t),\ \sigma(t)) \to (v_r(t),\ \sigma_r(t))$ as $t \to +\infty$ for all such solutions, but that theorem gives no information about bounds on the displacements corresponding to (v,σ). Blair has shown when $\mu/m=1$ that the product

$$\cos T_1\ \cos T_2 \ldots \cos T_n, \quad (35)$$

in which $T_1, T_2, \ldots T_n$ are the lengths of the subintervals during the period of (v_r,σ_r) on which (v_r,σ_r) satisfies the "elastic equations" $(29)_2$, (31), determines the limiting rate of convergence of (v,σ) to (v_r,σ_r). If this product is less than one in absolute value, then the displacements of (v,σ) are bounded if and only if those of (v_r,σ_r) are bounded. Blair's first result guarantees the boundedness of u_r, and his proof of that result shows that the product in (35) has absolute value less than 1. Hence, we conclude that every perturbation (v,σ) of (v_r,σ_r) also has bounded displacements.

Conclusion

I hope that this brief summary of recent research on elastic-plastic oscillators indicates not only the mathematical richness of this subject, but also the role that careful mathematical studies of these oscillators can play in understanding their physical properties.

References

[1] T. K. Caughey, Sinusoidal excitation of a system with bilinear hysteresis, J. Appl. Mech. **27**, Trans. ASME **82**, 640-643 (1960).

[2] M. A. Krasnoselskii et. al, Hysterant operator, Sov. Math. Dokl. **11**, 29-33 (1970).

[3] P. P. Zabreiko, M. A. Krasnoselskii, and E. A. Lifsic, An elastic-plastic element oscillator, Sov. Math. Dokl. **11**, 73-75 (1970).

[4] M. A. Krasnoselskii, A mathematical description of the oscillations of a material point on an elastic-plastic element, AMS Translation (2) **105**, 206-210 (1976).

[5] J. L. Buhite and D. R. Owen, An ordinary differential equation from the theory of plasticity, Arch. Rational Mech. Anal. **71**, 357-383 (1979).

[6] M. A. Krasnoselskii and A. V. Pokrovskii, Systems with Hysteresis (in Russian), Moskva, "Nauka" (1983).

[7] John P. Thomas, Ph.D. Thesis, Department of Mathematics, Carnegie Mellon University (1984).

[8] T. Miyoshi, Foundations of the Numerical Analysis of Plasticity, Lecture Notes in Numerical and Applied Analysis, Vol. 7, North Holland: Amsterdam, New York, Oxford (1985).

[9] D. R. Owen, Weakly decaying energy separation and uniqueness of motions of elastic-plastic oscillators with work-hardening, Arch. Rational Mech. Anal. **98**, 95-114 (1987).

[10] M. Mihaelescu-Suliciu, I. Siliciu, W. O. Williams, On viscoplastic and elastic-plastic oscillators, Q. Applied Math., to appear.

[11] D. R. Owen and J. P. Thomas, Conditions for Mechanical Self-Annealing in Motions of Elastic-Plastic Oscillators, Arch. Rational Mech. Anal., forthcoming.

[12] J. Bielak, D. R. Owen, and J. P. Thomas, Permanent drift of bilinear, hysteretic oscillators subject to impulsive loading.

[13] K. Gröger, J. Necas, and L. Trávnicek, Dynamic Deformation Processes of Elastic-Plastic Systems, ZAMM **59**, 567-572 (1979).

David R. Owen
Department of Mathematics
Carnegie Mellon University
Pittsburgh, PA 15213
U.S.A.

OPTIMAL CONTROL OF SYSTEMS GOVERNED BY VARIATIONAL-HEMIVARIATIONAL INEQUALITIES

P.D. Panagiotopoulos

1. Introduction

The present paper deals with the following optimal control problem: Minimize J(y,u) where (y u) are related by the variational hemivariational inequality

$$y\in V,\ (A(u)y,y^*-y)+\int_{\Omega'} j^0(y,y^*-y)\,d\Omega+\Phi(y^*)-\Phi(y)\geq$$
$$\geq (f+Bu,y^*-y)\qquad \forall y^*\in V.$$

Here the control u belongs to the admissible set U_{ad}, $A(u)\in\mathcal{L}(V,V')$, j is a locally Lipschitz continuous functional on R and Φ is a convex, l.s.c. and proper functional on V. The theory finds several applications in mechanics. They concern control and identification problems for laminated Kirchhoff and v.Kármán plates obeying nonmonotone, possibly multivalued interface laws (adhesive contact) and subjected to monotone possibly multivalued boundary conditions [1],[2].

There exists a variety of important mechanical problems involving nonmonotone, multivalued relations between stresses and strains, between reactions and displacements or generally between generalized forces and fluxes. Such relations expressed in terms of the new notion of nonconvex superpotential introduced in [3], [4], give rise to hemivariational inequalities. For their study we refer to [4]-[6]. The nonconvex superpotentials generalizes the notion of convex superpotential introduced in mechanics by J.J. Moreau [7]. Convex superpotentials describe monotone possibly multivalued mechanical laws and lead to variational inequalities. If in a problem both convex and nonconvex superpotentials appear, then we get as a variational formulation of it a variational hemivariational inequality. This new type of variational formulation has been studied in the articles of the author in [3],[4] as well as in [1],[2] and [8].

With respect to the hemivariational inequalities and to the variational hemivariational inequalities and the related mechanical problems the optimal control and identification problems have a great importance, not only because they give rise to a new type of mathematical problem, but also from the standpoint of mechanics because they permit a deeper study of the behaviour of the corresponding physical problem.

The resulting mathematical problem is of nonclassical nature, because, instead of state equations we have state hemivariational or variational hemivariational inequalities. We recall here that optimal control problems having state variational inequalities have been already formulated and studied (cf. e.g. [9]-[16]). If the state variables are connected with the control variables through a hemivariational or a variational hemivariational inequality, monotonicity arguments cannot be used and the application of compactness arguments becomes necessary.

Optimal control and identification problems of systems governed by hemivariational inequalities have been already studied by the author and J. Haslinger in [17],[18]. Here we shall study the respective problems but with a coercive variational-hemivariational inequality as a state relation.

Let V be a Hilbert space, V' its dual space and $(.,.)$ the duality pairing. We assume that V is defined on an open bounded subset $\Omega\subset R^n$ and let us assume that $V\subset L^2(\Omega)\subset V'$, where the injections are continuous, that

$$V \subset L^{\infty}(\Omega) \quad \text{is compact} \tag{1.1}$$

and that

$$V \cap L^{\infty}(\Omega') \quad \text{is dense in } V \text{ for the } V\text{-norm} \tag{1.2}$$

where $\Omega' \subset\subset \Omega$. We consider the following problem for $f \in V'$ and for $\Phi: V \to (-\infty, +\infty]$, $\Phi \not\equiv \infty$, convex and l.s.c. functional.

Find $y \in V$ such that

$$a(y, y^* - y) + \int_{\Omega'} j^0(y, y^* - y)\, d\Omega + \Phi(y^*) - \Phi(y) \geq$$

$$\geq (f, y^* - y) \quad \forall y \in V . \tag{1.3}$$

Here $a(.,.)$ denotes a bilinear symmetric continuous coercive form on $V \times V$, j is a nonconvex superpotential and j^0 is the directional derivative of F.H. Clarke [19]. In order to formulate the corresponding optimal control problem we introduce a control variable u through an operator B, i.e. we replace f by $f + Bu$, assuming that the energy of the system depends also on u, and we want to determine u and y which minimize an appropriately defined cost functional and satisfy the control-state hemivariational inequality (1.3). Note that (1.3) corresponds to several important mechanical problems (cf. [1],[2],[8]).

2. Mathematical Formulation of the Problem

Let $b \in L^{\infty}_{loc}(R)$. For any $\mu > 0$ we define

$$\underline{b}_{\mu}(\xi) = \operatorname*{ess\,inf}_{|t-\xi| \leq \mu} b(t), \quad \overline{b}_{\mu}(\xi) = \operatorname*{ess\,sup}_{|t-\xi| \leq \mu} b(t) \tag{2.1}$$

$$\underline{b}(\xi) = \lim_{\mu \to 0_+} \underline{b}_{\mu}(\xi) \quad , \quad \overline{b}(\xi) = \lim_{\mu \to 0_+} \overline{b}_{\mu}(\xi) . \tag{2.2}$$

Then $\hat{b}(\xi) = [\underline{b}(\xi), \overline{b}(\xi)]$ is a multivalued function on R resulting from b by "filling in the jumps" and let

$$j(\xi) = \int_0^{\xi} b(t)dt \tag{2.3}$$

a locally Lipschitz continuous function. It is proved in [20], that the generalized derivative of j at ξ (in the sense of F.H. Clarke) satisfies $\overline{\partial} j(\xi) \subseteq \hat{b}(\xi)$ and if $b(\xi_{\pm 0})$ exists at any $\xi \in R$ then

$$\overline{\partial} j(\xi) = \hat{b}(\xi) . \tag{2.4}$$

Let us assume the variational hemivariational inequality of the following type:

$$(\mathcal{P}(u)) \begin{cases} \text{Find } y = y(u) \in V \text{ such that} \\ (A(u)y, y^*-y) + \int_{\Omega'} j^0(y, y^*-y)\,d\Omega + \Phi(y^*) - \Phi(y) \geqq \\ \geqq (f + Bu, y^*-y) \quad \forall y^* \in V . \end{cases} \tag{2.5}$$

Here we assume that the control variable $u \in U_{ad}$, where U_{ad} is a nonempty, closed, convex subset of a Hilbert space $U \subset V'$, $B \in \mathcal{L}(U, V')$ (i.e. a linear, compact mapping), $f \in V'$ a given fixed element and Φ is a convex, l.s.c. and proper functional on V. We assume that for any $u \in U_{ad}$, $A(u) \in \mathcal{L}(V, V')$ is generated by a bilinear, continuous form

$$(A(u)y, y^*) = a_u(y, y^*) \quad \forall y, y^* \in V , \tag{2.6}$$

which satisfies:

$$\exists c \text{ const} > 0: \ a_u(y,y) \geqq c||y||^2 \quad \forall y \in V \quad \forall u \in U_{ad} ; \tag{2.7}$$

$$a_u(y, y^*) = a_u(y^*, y) \quad \forall y, y^* \in V \quad \forall u \in U_{ad} ; \tag{2.8}$$

$$u_n \to u \text{ weakly in } U \Rightarrow A(u_n) \to A(u) \text{ in } \mathcal{L}(V, V') . \tag{2.9}$$

Definition 2.1. A function $y = y(u) \in V$ is called a solution of $(\mathcal{P}(u))$ if there exists $\chi \in L^1(\Omega') \cap V'$ such that (cf. [1], [5], [6])

$$(A(u)y,y^*) + (\chi,y^*)|_{\Omega'} + \Phi(y^*) - \Phi(y) \geqq$$

$$\geqq (f + Bu,y^*) \quad \forall y^* \in V \tag{2.10}$$

$$\chi(x) \in \hat{b}(y(x)) \quad \text{a.e. in } \Omega' . \tag{2.11}$$

Finally, let $J:V \times U \to R$ be a cost functional, satisfying
J is coercive in u, uniformly with respect to y, i.e.

$$\forall k > 0 \quad \exists \tau > 0 \quad \forall ||u|| \geqq \tau,\ u \in U_{ad},\quad \forall y \in V : J(y,u) \geqq k; \tag{2.12}$$

$$\left.\begin{array}{l} y_n \to y \text{ (weakly) in } V \\ u_n \to u \text{ weakly in } U \end{array}\right\} \Rightarrow \lim_{n\to\infty} J(y_n,u_n) = J(y,u) . \tag{2.13}$$

The optimal control problem for the variational hemi-variational inequality ($\mathcal{P}(u)$) reads:

$$\text{(P)} \quad \left\{ \begin{array}{l} \text{Find } \bar{u} \in U_{ad} \text{ such that} \\ J(y(\bar{u}),\bar{u}) \leqq J(y(u),u) \quad \forall u \in U_{ad} , \end{array} \right. \tag{2.15}$$

where $y(u) \in V$ and $u \in U_{ad}$ are related by ($\mathcal{P}(u)$).

In order to show the existence of a solution of (P) we shall make use of the following assumption concerning b:
$\exists \bar{\xi} > 0$ such that

$$\operatorname*{ess\,sup}_{\xi \in (-\infty,-\bar{\xi})} b(\xi) \leqq 0 \leqq \operatorname*{ess\,inf}_{\xi \in (\bar{\xi},+\infty)} b(\xi) . \tag{2.16}$$

We shall distinguish two cases: the differentiable case and the nondifferentiable case. In the first (resp. second) case we assume that $\text{grad}\Phi(\cdot)$ exists everywhere (resp. does not exist everywhere). Note that in the differentiable case ($\mathcal{P}(u)$) takes the following equivalent form:

$$\begin{cases} \text{Find } y = y(u) \in V \text{ such that} \\ (A(u)y, y^*-y) + \int_{\Omega'} j^0(y, y^*-y)\,d\Omega + (\operatorname{grad}\Phi(y), y^*-y) \geq \end{cases}$$

$$\geq (f + Bu, y^*-y) \quad \forall y^* \in V \tag{2.17}$$

The nondifferentiable case is treated by regularization of Φ i.e. we assume that there exist Gâteaux differentiable functionals $\Phi_\rho, \rho>0$, which have the following properties

$$\text{(i)} \quad \Phi_\rho(y^*) \to \Phi(y^*) \quad \forall y^* \in V \quad \text{as } \rho \to 0 . \tag{2.18}$$

$$\text{(ii)} \quad \operatorname{grad}\Phi_\rho(0) = 0 \quad \text{for every } \rho . \tag{2.19}$$

(iii) If $y_\rho \to y$ weakly in V as $\rho \to 0$ and

$\Phi_\rho(y_\rho) < M$ where M is a constant, then

$$\liminf_{\rho \to 0} \Phi_\rho(y_\rho) \geq \Phi(y) . \tag{2.20}$$

Therefore we shall symbolize the differentiable problems as $(\mathcal{P}_\rho(u))$ and the corresponding optimal control problem as (P_ρ).

3. The Finite Dimensional Problem $(P_\rho)_h$

Let $V_h, U_h, h \in (0,1)$, be finite-dimensional subspaces of $V \cap L^\infty(\Omega')$ and U respectively. Let U^h_{ad} be a closed, convex subset of U_h, not necessarily contained in U_{ad}. With respect to $(\mathcal{P}_\rho(u))$ we consider its discretized form $(u_h \in U^h_{ad})$

$$(\mathcal{P}_\rho(u_h))_h \begin{cases} \text{Find } y_h = y_h(u_h) \in V_h \text{ such that} \\ (A(u_h)y_h, y_h^*) + \int_{\Omega} \chi_h^\rho y_h^* d\Omega + (\operatorname{grad}\Phi_\rho(y_h^\rho), y_h^*) = \\ = (f + Bu_h^\rho, y_h^*) \quad \forall y_h^* \in V_h \qquad (3.1) \\ \chi_h^\rho \in L^1(\Omega') \cap V' \quad \text{and} \\ \chi_h^\rho(x) \in \hat{b}(y_h^\rho(x)) \quad \text{a.e. in } \Omega' . \qquad (3.2) \end{cases}$$

We now define the finite dimensional optimal control problem

$$(P_\rho)_h \qquad \begin{array}{l} \text{Find } \bar{u}_h^\rho \in U_{ad}^h \text{ such that} \\ J(y_h^\rho(\bar{u}_h^\rho),\bar{u}_h^\rho) \leq J(y_h^\rho(u_h),u_h) \quad \forall u_h \in U_{ad}^h, \end{array} \tag{3.3}$$

where $y_h^\rho(u_h)$ is a solution of $(\mathcal{P}_\rho(u_h))_h$. Further we shall prove the existence of at least one solution of $(P_\rho)_h$. But first we shall show that $(\mathcal{P}_\rho(u_h))_h$ has at least one solution for any $u_h \in U_{ad}^h$. We suppose that (2.7) holds not only for $u \in U_{ad}$, but also for $u \in \bigcup_h U_{ad}^h$.

Let $\omega \in \mathcal{D}(R)$ be a mollifier, i.e. $\int_{-\infty}^{+\infty} \omega(\xi)d\xi = 1$, $\omega \geq 0$ $\omega_\varepsilon(\xi) = \frac{1}{\varepsilon}\omega(\frac{\xi}{\varepsilon})$ and let

$$b_\varepsilon(\xi) = \int_{-\infty}^{+\infty} \omega_\varepsilon(t)b(\xi - t)dt = b * \omega_\varepsilon \quad (\varepsilon > 0) . \tag{3.4}$$

Lemma 3.1. Suppose that (2.16) and (2.19) hold. Then for any $u_h \in U_{ad}^h$ there exists at least one solution of the problem:

$$\begin{array}{l} \text{Find } y_{h\varepsilon}^\rho = y_{h\varepsilon}^\rho(u_h) \in V_h \text{ such that} \\ (A(u_h)y_{h\varepsilon}^\rho, y_h^*) + \int_{\Omega'} b_\varepsilon(y_{h\varepsilon}^\rho)y_h^* d\Omega + (\mathrm{grad}\Phi_\rho(y_{h\varepsilon}^\rho), y_h^*) = \\ = (f + Bu_h, y_h^*) \quad \forall y_h^* \in V_h . \end{array} \tag{3.5}$$

Proof: From the assumption (2.16) we get that there exist positive numbers ρ_1, ρ_2 such that for any $\varepsilon > 0$

$$\begin{array}{lll} |b_\varepsilon(\xi)| \leq \rho_2 & \text{for} & |\xi| \leq \rho_1 \\ b_\varepsilon(\xi)| \geq 0 & \text{for} & \xi \geq \rho_1 \\ b_\varepsilon(\xi) \leq 0 & \text{for} & \xi \leq -\rho_1 . \end{array} \tag{3.6}$$

For the mapping $T: V_h \to V_h'$

$$(Tz_h,\cdot) = (A(u_h)z_h,\cdot) + \int_{\Omega'} b_\varepsilon(z_h)\cdot d\Omega + (\mathrm{grad}\Phi_\rho(z_h),\cdot)$$
$$-(f + Bu_h,\cdot) \quad , \tag{3.7}$$

we can verify that for u_h given we have

$$(Tz_h,z_h) \geqq c||z_h||^2 - \rho_1\rho_2 \mathrm{mes}\Omega' - c||z_h||, \tag{3.8}$$

because due to the monotonicity of $\mathrm{grad}\Phi_\rho$ and (2.19) we can write that

$$(\mathrm{grad}\Phi_\rho(z_h),z_h) \geqq 0 \quad \forall z_h \in V_h \tag{3.9}$$

and using the fact that $A(\cdot)$ is V-elliptic for $u \in U_{ad} \cup (\bigcup_h U^h_{ad})$.

From (3.8) by applying Brouwer's fixed point theorem we prove the lemma. q.e.d.

<u>Proposition 3.1.</u> Let (2.16) and (2.19) hold. For any $u_h \in U^h_{ad}$ there exists at least one solution $y^\rho_h \in V_h$ of $(\mathscr{P}_\rho(u_h))_h$.

Proof: Let $y^\rho_{h\varepsilon} \in V_h$ be a solution of (3.5) and let $\varepsilon \to 0_+$. From Brouwer's theorem we easily see that $\{y^\rho_{h\varepsilon}\}$ is bounded independently of ε,ρ and h. Thus

$$y^\rho_{h\varepsilon} \to y^\rho_h \in V_h, \quad \varepsilon \to 0_+ . \tag{3.10}$$

At the same time by setting $y^*_h = y^\rho_{h\varepsilon}$ we get from (3.5)

$$|b_\varepsilon(y^\rho_{h\varepsilon})|_{L^2} \leqq c(h) \quad . \tag{3.11}$$

Thus we obtain that there exists a function $\chi_h \in L^2(\Omega')$, such that

$$b_\varepsilon(y^\rho_{h\varepsilon}) \to \chi_h \quad \text{(weakly) in } L^2(\Omega'), \quad \varepsilon \to 0_+ \quad . \tag{3.12}$$

Moreover (3.5) implies that there exist c independent of ε such that

$$||\text{grad}\Phi_\rho(y^\rho_{h\varepsilon})||_{V'} \leq c \; . \tag{3.13}$$

Thus as $\varepsilon \to 0_+$

$$\text{grad}\Phi_\rho(y^\rho_{h\varepsilon}) \to \psi^\rho_h \quad \text{(weakly) in } V'. \tag{3.14}$$

Passing to the limit with $\varepsilon \to 0_+$ in (3.5) we obtain

$$(A(u_h)y^\rho_h, y^*_h) + \int_{\Omega'} \chi^\rho_h y^*_h d\Omega + (\psi^\rho_h, y^*_h) = (f + Bu_h, y^*_h) \quad \forall y^*_h \in V_h \; . \tag{3.15}$$

It remains to show that

$$\psi^\rho_h = \text{grad}\Phi_\rho(y^\rho_h) \quad \text{in } V' \tag{3.16}$$

and that

$$\chi^\rho_h \in \hat{b}(y^\rho_h) \quad \text{a.e.} \quad \text{in } \Omega'. \tag{3.17}$$

Let us show first (3.16). Due to the monotonicity of $\text{grad}\Phi_\rho(\cdot)$ we have that

$$X^\rho_{h\varepsilon} = (\text{grad}\Phi(y^\rho_{h\varepsilon}) - \text{grad}\Phi(\Theta), y^\rho_{h\varepsilon} - \Theta) \geq 0 \qquad \forall \Theta \in V \; . \tag{3.18}$$

By means of (3.5) we get

$$X^\rho_{h\varepsilon} = (f + Bu_h, y^\rho_{h\varepsilon}) - \int_{\Omega'} b_\varepsilon(y^\rho_{h\varepsilon}) y^\rho_{h\varepsilon} d\Omega - (A(u_h)y^\rho_{h\varepsilon}, y^\rho_{h\varepsilon}) - (\text{grad}\Phi(\Theta), y^\rho_{h\varepsilon} - \Theta) - (\text{grad}\Phi(y^\rho_{h\varepsilon}), \Theta) \geq 0. \tag{3.19}$$

We can verify that (cf. assumption (1.1))

$$\int_{\Omega'} b_\varepsilon(y^\rho_{h\varepsilon}) y^\rho_{h\varepsilon} d\Omega \to \int_{\Omega'} \chi^\rho_h y^\rho_h d\Omega \; . \tag{3.20}$$

Indeed

$$\int_{\Omega'} b_\varepsilon(y^\rho_{h\varepsilon})y^\rho_{h\varepsilon}d\Omega - \int_{\Omega'} \chi^\rho_h y^\rho_h d\Omega =$$

$$= \int_{\Omega'} b_\varepsilon(y^\rho_{h\varepsilon})(y^\rho_{h\varepsilon}-y^\rho_h)d\Omega + \int_{\Omega'} y^\rho_h(b_\varepsilon(y^\rho_{h\varepsilon})-\chi^\rho_h)d\Omega = A+B. \tag{3.21}$$

Because of (3.10) $A \to 0$ and because of (3.12) $B \to 0$. Thus from (3.19) we obtain for $\varepsilon \to 0$

$$0 \leq \limsup X^\rho_{h\varepsilon} \leq -(A(u_h)y^\rho_h, y^\rho_h) - \int_{\Omega'} \chi^\rho_h y^\rho_h d\Omega + (f + Bu_h, y^\rho_h) -$$

$$-(\psi^\rho_h, \Theta) - (\mathrm{grad}\Phi(\Theta), y^\rho_h - \Theta) . \tag{3.22}$$

From (3.22) and (3.15) by setting $y^*_h = y^\rho_h$ we obtain

$$(\psi^\rho_h - \mathrm{grad}\Phi(\Theta), y^\rho_h - \Theta) \geqq 0 \quad \forall \Theta \in V \tag{3.23}$$

which implies by Minty's monotonicity argument that (3.16) holds. Indeed setting in (3.23) $y^\rho_h - \Theta = \lambda w$, $\lambda > 0$, we get the expression

$$(\psi^\rho_h - \mathrm{grad}\Phi(y^\rho_h - \lambda w), w) \geqq 0 \quad \forall w \in V . \tag{3.24}$$

But $\lambda \to \mathrm{grad}\Phi(y^\rho_h - \lambda w)$ is monotone and thus we can take $\lambda \to 0_+$ in (3.24). Thus

$$(\psi^\rho_h - \mathrm{grad}\Phi(y^\rho_h), w) \geqq 0 \quad \forall w \in V . \tag{3.25}$$

Setting $\pm w$ in (3.25) we get (3.16). It remains to show (3.17). Indeed, starting from the inequality

$$\operatorname*{ess\,inf}_{|t| \leq \varepsilon} b(\xi - t) \leq b_\varepsilon(\xi) \leq \operatorname*{ess\,sup}_{|t| \leq \varepsilon} b(\xi - t),$$

which is verified using the definition of b_ε, we find that

$$\operatorname*{ess\,inf}_{|\xi - y^\rho_{h\varepsilon}| \leq \varepsilon} b(\xi) \leq b_\varepsilon(y^\rho_{h\varepsilon}) \leq \operatorname*{ess\,sup}_{|\xi - y^\rho_{h\varepsilon}| \leq \varepsilon} b(\xi) . \tag{3.26}$$

Let $\mu>0$ be given and $0<\varepsilon<\frac{\mu}{2}$. Then

$$\operatorname*{ess\,inf}_{|\xi-y^\rho_{h\varepsilon}|\le\frac{\mu}{2}} b(\xi) \le b_\varepsilon(y^\rho_{h\varepsilon}) \le \operatorname*{ess\,sup}_{|\xi-y^\rho_{h\varepsilon}|\le\frac{\mu}{2}} b(\xi) \; . \tag{3.27}$$

From (3.27) and (3.10) we obtain

$$\underline{b}_\mu(y^\rho_h) = \operatorname*{ess\,inf}_{|\xi-y^\rho_h|\le\mu} b(\xi) \le b_\varepsilon(y^\rho_{h\varepsilon}) \le \operatorname*{ess\,sup}_{|\xi-y^\rho_h|\le\mu} b(\xi) = \overline{b}_\mu(y^\rho_h). \tag{3.28}$$

Multiplying (3.28) by $\varphi\in L^2(\Omega')$, $\varphi\ge 0$ a.e. on Ω', and using (3.12) we have

$$\int_{\Omega'} \underline{b}_\mu(y^\rho_h)\varphi d\Omega \le \int_{\Omega'} \chi^\rho_h \varphi d\Omega \le \int_{\Omega'} \overline{b}_\mu(y^\rho_h)\varphi d\Omega \; . \tag{3.29}$$

Finally, setting $\mu\to 0_+$ in (3.29) we obtain

$$\int_{\Omega'} \underline{b}(y^\rho_h)\varphi d\Omega \le \int_{\Omega'} \chi^\rho_h \varphi d\Omega \le \int_{\Omega'} \overline{b}(y^\rho_h)\varphi d\Omega \; , \tag{3.30}$$

which implies (3.17). q.e.d.

Now we prove the existence of a solution of the finite dimensional optimal control problem $(P_\rho)_h$.

<u>Proposition 3.2.</u> There exists at least one solution of $(P_\rho)_h$.

Proof: Set

$$q = \inf_{u^\rho_h\in U_{ad}} J(y^\rho_h, u^\rho_h) \tag{3.31}$$

and let $y^{\rho n}_h, u^{\rho n}_h$ be a minimizing sequence, i.e.

$$q = \lim_{n\to\infty} J(y^{\rho n}_h, u^{\rho n}_h) \; , \tag{3.32}$$

where $y^{\rho n}_h$ is a solution of $(\mathcal{P}_\rho(u^{\rho n}_h))_h$.

Because of (2.12) and (3.10), $\{u_h^{\rho n}\}$ and also $\{y_h^{\rho n}\}$ are bounded independently of n. Without loss of generality we may write that as $n \to \infty$

$$\begin{aligned} y_h^{\rho n} &\to \bar{y}_h^{\rho} \\ & \qquad\qquad n \to \infty . \\ u_h^{\rho n} &\to \bar{u}_h^{\rho} , \end{aligned} \tag{3.33}$$

But $y_h^{\rho n}, u_h^{\rho n}$ are related by

$$(A(u_h^{\rho n}) y_h^{\rho n}, y_h^*) + \int_{\Omega'} \chi_h^{\rho n} y_h^* d\Omega + (\mathrm{grad}\Phi_\rho(y_h^{\rho n}), y_n^*) =$$

$$= (f + Bu_h^{\rho n}, y_h^*) \quad \forall y_h^* \in V_h, \quad \chi_h^{\rho n} \in \hat{b}(y_h^{\rho n}) \quad \text{a.e. in } \Omega'. \tag{3.34}$$

Due to

$$\underline{b}(y_h^{\rho n}) \leq \chi_h^{\rho n} \leq \bar{b}(y_h^{\rho n}) \quad \text{a.e. in } \Omega' \tag{3.35}$$

there exists $\chi_h^{\rho} \in L^2(\Omega')$ such that as $n \to \infty$

$$\chi_h^{\rho n} \to \chi_h^{\rho} \quad \text{weakly in } L^2(\Omega'). \tag{3.36}$$

Moreover we obtain from (3.34) that

$$\| \mathrm{grad}\Phi_\rho(y_h^{\rho n}) \|_{V'} < c \tag{3.37}$$

independently of n and thus

$$\mathrm{grad}\Phi_\rho(y_h^{\rho n}) \to \psi_h^{\rho} \quad \text{weakly in } V'. \tag{3.38}$$

Passing to the limit $n \to \infty$ in (3.34) and using (2.9) we arrive at

$$A(\bar{u}_h^{\rho}) \bar{y}_h^{\rho}, y_h^*) + \int_{\Omega'} \chi_h^{\rho} y_h^* d\Omega + (\psi_h^{\rho}, y_h^*) = (f + B\bar{u}_h^{\rho}, y_h^*)$$

$$\forall y_h^* \in V_h . \tag{3.39}$$

It remains to show that

$$\psi_h^\rho = \text{grad}\Phi_\rho(\overline{y}_h^\rho) \text{ in } V' \tag{3.40}$$

and

$$\chi_h^\rho \in \hat{b}(\overline{y}_h^\rho) \quad \text{a.e. in } \Omega' . \tag{3.41}$$

The proof of (3.40) is the same as the proof of (3.16) i.e. it is achieved by Minty's monotonicity argument. For the proof of (3.41) we use the fact that $\underline{b}$ (resp. $\overline{b}$) is lower (resp. upper) semicontinuous function, respectively.

Let $\varphi \in L^2(\Omega'), \varphi \geq 0$ a.e. in Ω'. Then

$$\begin{aligned}\int_{\Omega'} \chi_h^\rho \varphi d\Omega = \lim_{n\to\infty} \int_{\Omega'} \chi_h^{\rho n} \varphi d\Omega \leq \limsup_{n\to\infty} \int_{\Omega'} \overline{b}(y_h^{\rho n})\varphi d\Omega \\ \leq \int_{\Omega'} \limsup \overline{b}(y_h^{\rho n})\varphi d\Omega \leq \int_{\Omega'} \overline{b}(\overline{y}_h^\rho)\varphi d\Omega .\end{aligned} \tag{3.42}$$

On the other hand

$$\begin{aligned}\lim_{n\to\infty} \int_{\Omega'} \chi_h^{\rho n} \varphi d\Omega \geq \liminf_{n\to\infty} \int_{\Omega'} \underline{b}(y_h^{\rho n})\varphi d\Omega \\ \geq \int_{\Omega'} \liminf \underline{b}(y_h^{\rho n})\varphi d\Omega \geq \int_{\Omega'} \underline{b}(\overline{y}_h^\rho)\varphi d\Omega .\end{aligned} \tag{3.43}$$

Relations (3.42) and (3.43) imply (3.41); i.e. $\overline{y}_h$ is a solution of $(\mathcal{P}(\overline{u}_h))_h$. Finally from (2.13) and (3.32) we obtain that

$$q = J(\overline{y}_h^\rho, \overline{u}_h^\rho) . \tag{3.44}$$

q.e.d.

4. Existence of Solutions of the Continuous Problem (P)

In this Section we show that (P) has at least one solution.

Theorem 4.1. Suppose that (2.16),(2.18)÷(2.20) hold. (P) has at least one solution.

Proof: If (2.16)÷(2.20) hold, the state problem $(\mathcal{P}_\rho(u))$ has at least one solution $y^\rho = y(u^\rho)$ for any $u^\rho \in U_{ad}$ ([1],[5],[6]). Let $\{y^\rho_{(n)}, u^\rho_{(n)}\}$ be a minimizing sequence, i.e.

$$q \equiv \inf_{u^\rho \in U_{ad}} J(y^\rho(u^\rho), u^\rho) = \lim_{n\to\infty} J(y^\rho_{(n)}, u^\rho_{(n)}). \tag{4.1}$$

$y^\rho_{(n)}$ and $u^\rho_{(n)}$ are related by

$$(A(u^\rho_{(n)})y^\rho_{(n)}, y^*) + (\chi^\rho_{(n)}, y^*)|_{\Omega'} + (\operatorname{grad}\Phi_\rho(y^\rho_{(n)}), y^*) =$$

$$= (f + Bu^\rho_{(n)}, y^*) \quad \forall y^* \in V ; \tag{4.2}$$

$$\chi^\rho_{(n)} \in \hat{b}(y^\rho_{(n)}) \text{ a.e. on } \Omega' \quad \text{for all n.} \tag{4.3}$$

We note also here the monotonicity inequality

$$(\operatorname{grad}\Phi_\rho(y^\rho_{(n)}), y^\rho_{(n)}) \geqq 0 . \tag{4.4}$$

We obtain now from (4.2),(4.3) and (4.4), using (2.12) and (2.16), that the sequences $\{y^\rho_{(n)}\}\{u^\rho_{(n)}\}$ are bounded in the corresponding spaces independently of ρ,n. Without loss of generality we may write that as $n\to\infty$

$$y^\rho_{(n)} \to \bar{y}^\rho \text{ weakly in } V \tag{4.5}$$

$$u^\rho_{(n)} \to \bar{u}^\rho \text{ weakly in } U_{ad} \tag{4.6}$$

and

$$y^\rho_{(n)} \to \bar{y}^\rho \text{ in } L^\infty(\Omega'), \tag{4.7}$$

making use of (1.1). From the relations (4.3) and (1.1) we obtain that $\{\chi^\rho_{(n)}\}$ is bounded in $L^2(\Omega')$-norm. Again we obtain that

$$\chi^{\rho}_{(n)} \to \bar{\chi}^{\rho} \text{ weakly in } L^2(\Omega'). \tag{4.8}$$

From (4.3) and the above relations we obtain that $\mathrm{grad}\Phi_{\rho}(y^{\rho}_{(n)})$ is bounded in V' and thus we may write that

$$\mathrm{grad}\Phi_{\rho}(y^{\rho}_{(n)}) \to \bar{\Psi}^{\rho} \text{ weakly in } V' . \tag{4.9}$$

Passing to the limit with $n\to\infty$ and using (2.9),(4.5)÷(4.9) we arrive at

$$(A(\bar{u}^{\rho})\bar{y}^{\rho},y^*) + (\bar{\chi}^{\rho},y^*)|_{\Omega'} + (\bar{\Psi}^{\rho},y^*) =$$

$$= (f + B\bar{u}^{\rho},y^*) \quad \forall y^*\in V . \tag{4.10}$$

It remains to show that

$$\bar{\chi}^{\rho}\in\hat{b}(\bar{y}^{\rho}) \tag{4.11}$$

and that

$$\bar{\Psi}^{\rho}=\mathrm{grad}\Phi_{\rho}(\bar{y}^{\rho}) . \tag{4.12}$$

To prove (4.11) we use the same approach as in the proof of Proposition 3.2.

Let $\varphi\in L^2(\Omega')$, $\varphi \geq 0$ a.e. on Ω'. Then

$$\begin{aligned}\int_{\Omega'}\bar{\chi}^{\rho}\varphi d\Omega &= \lim_{n\to\infty} \int \chi^{\rho}_{n}\varphi d\Omega \leq \\ &\leq \limsup_{n\to\infty} \int_{\Omega'} \bar{b}(y^{\rho}_{n})\varphi d\Omega \leq \\ &\leq \int_{\Omega'} \limsup_{n\to\infty} \bar{b}(y^{\rho}_{n})\varphi d\Omega \leq \\ &\leq \int_{\Omega'} \bar{b}(\bar{y}^{\rho})\varphi d\Omega \end{aligned} \tag{4.13}$$

making use of (4.7),(4.8) and the upper semicontinuity of $\bar{b}$.

The analogous inequality

$$\int_{\Omega'} \overline{\chi}^{\rho} \varphi d\Omega \geq \int_{\Omega\Gamma} \underline{b}(\overline{y}^{\rho}) \varphi d\Omega \tag{4.14}$$

can be proved in the same way.

It remains to show (4.12). Due to the monotonicity of $\text{grad}\Phi_{\rho}(\cdot)$ we have that

$$X_n^{\rho} = (\text{grad}\Phi(y_{(n)}^{\rho}) - \text{grad}\Phi(\Theta), y_{(n)}^{\rho} - \Theta) \geq 0 \quad \forall\Theta\in V \tag{4.15}$$

and by means of (4.3) we have

$$0 \leq X_n^{\rho} = (f + Bu_{(n)}^{\rho}, y_{(n)}^{\rho}) - (\chi_{(n)}^{\rho}, y_n^{\rho})|_{\Omega'} - (A(u_n^{\rho}) y_n^{\rho}, y_n^{\rho}) -$$

$$- (\text{grad}\Phi(\Theta), y_{(n)}^{\rho} - \Theta) - (\text{grad}\Phi(y_{(n)}^{\rho}), \Theta). \tag{4.16}$$

But

$$(\chi_{(n)}^{\rho}, y_{(n)}^{\rho})|_{\Omega'} \to (\overline{\chi}^{\rho}, \overline{y}^{\rho})|_{\Omega'} . \tag{4.17}$$

Indeed the difference of the two quantities in (4.17) is equal to

$$(\chi_{(n)}^{\rho}, (y_n^{\rho} - \overline{y}^{\rho}))|_{\Omega'} + (\overline{y}^{\rho}, (\chi_{(n)}^{\rho} - \overline{\chi}^{\rho}))|_{\Omega'} \tag{4.18}$$

which tends to zero because of (4.7) and (4.8). Thus we have, using the compactness of B, (2.7),(2.9) and (4.17), that

$$0 \leq \limsup X_n^{\rho} \leq -(A(\overline{u}^{\rho})\overline{y}^{\rho}, \overline{y}^{\rho}) - (\overline{\chi}^{\rho}, \overline{y}^{\rho})|_{\Omega'} + (f + B\overline{u}^{\rho}, \overline{y}^{\rho}) -$$

$$- (\overline{\psi}^{\rho}, \Theta) - (\text{grad}\Phi(\Theta), \overline{y}^{\rho} - \Theta) \tag{4.19}$$

which together with (4.10) implies that

$$(\overline{\psi}^{\rho} - \text{grad}\Phi^{\rho}(\Theta), \overline{y}^{\rho} - \Theta) \geq 0 \quad \forall\Theta\in V . \tag{4.20}$$

Minty's theorem implies (4.12).

Combining (4.11),(4.12) with (4.10) we obtain that $\overline{y}^\rho$ solves the problem $(P_\rho(\overline{u}_\rho))$. Finally (4.1),(4.5) and (4.6) give with (2.13) that

$$q = J(\overline{y}^\rho,\overline{u}^\rho) \tag{4.21}$$

i.e. that $\{\overline{y}^\rho,\overline{u}^\rho\}$ is an optimal pair of $(\mathbb{P}_\rho)$. In the sequel we omit the overbar on $\overline{y}^\rho$ and $\overline{u}^\rho$.

To complete the proof we shall pass to the limit $\rho\to 0$. Again we have (cf. (4.5) etc.) that as $\rho\to 0$

$$u^\rho \to \overline{u} \quad \text{weakly in } U_{ad} \tag{4.22}$$

$$y^\rho \to \overline{y} \quad \text{weakly in V and strongly in } L^\infty(\Omega') \tag{4.23}$$

$$\chi^\rho \to \overline{\chi} \quad \text{weakly in } L^2(\Omega')\ . \tag{4.24}$$

From (4.10) and (4.12) we obtain that

$$(A(u^\rho)y^\rho,y^*-y^\rho) + (\chi^\rho,y^*-y^\rho)|_{\Omega'} + \Phi_\rho(y^*)-\Phi_\rho(y^\rho) \geqq$$

$$\geqq (f + Bu^\rho,y^*-y\rho) \qquad \forall y^*\in V\ . \tag{4.25}$$

We choose in (4.25) a y^* such that $\Phi(y^*)<\infty$. Then (2.18) implies that $\Phi_\rho(y^*)<M$ where M is a constant and thus we derive from (4.25) that $\Phi_\rho(y^\rho)<c$. Accordingly (2.20) holds. Now from (4.25) we get

$$D = \Phi_\rho(y^*) + (A(u^\rho)y^\rho,y^*) + (\chi^\rho,y^*)|_{\Omega'}-(f + Bu^\rho,y^*) \geqq$$

$$\geqq \Phi_\rho(y^\rho) + (A(u^\rho)y^\rho,y^\rho) + (\chi^\rho,y^\rho)|_{\Omega'}-(f + Bu^\rho,y^\rho) = F. \tag{4.26}$$

Thus as $\rho\to 0$ we may write by taking into account (2.9),(2.18), (2.20), the compactness of B and (4.22)÷(4.24) that

$$\liminf D = \lim\{\Phi_\rho(y^*)+(A(u^\rho)y^\rho,y^*)+(\chi^\rho,y^*)|_{\Omega'}-$$
$$-(f+Bu^\rho,y^*)\} = \Phi(y^*)+(A(\bar{u})\bar{y},y^*)+(\bar{\chi},y^*)|_{\Omega'}-$$
$$-(f+B\bar{u},y^*) \geqq \liminf F \geqq \Phi(y)+(A(\bar{u})\bar{y},\bar{y})+$$
$$+(\bar{\chi},\bar{y})|_{\Omega'}-(f+B\bar{u},\bar{y}) \quad \forall y^*\in V . \tag{4.27}$$

Here we have used also the relation

$$(\chi^\rho,y^\rho)|_{\Omega'} \to (\bar{\chi},\bar{y}) \text{ as } \rho\to 0 . \tag{4.28}$$

Its proof is similar to the proof of (4.17). Moreover we have used the inequality

$$\liminf (A(u^\rho)y^\rho,y^\rho) \geqq (A(\bar{u})\bar{y},\bar{y}) \tag{4.29}$$

which holds due to (2.7)÷(2.9). From (4.27) and from (4.22), (4.23) and (2.13) we obtain that $\{\bar{y},\bar{u}\}$ is an optimal pair of the optimal control problem ($\mathbb{P}$) q.e.d.

5. Applications

We close this paper by giving an application concerning control of problems governed by variational-hemivariational inequalities.

Let us consider the parameter identification problem for a rock with given interfaces. We assume that the behaviour of each interface is described by nonmonotone multivalued laws in the tangential and in the normal directions and that the rock obeys a convex superpotential law (i.e. a monotone multivalued stress-strain law of the form $\sigma \in \partial w(\varepsilon)$; ∂ is the subdifferential, w is convex, l.s.c. and proper). The form of the interface laws is not completely known. However, if measurements of some variables, i.e. of some displacements, are available, then a fictitious interface law given by a set of control parameters can be introduced and then the discrepancy between the measured values of the displacements and the predicted ones can be minimized. The same problem can be posed if the rock mass is completely inhomogeneous and we want to determine one or possibly several fictitious moduli of elasticity, which will minimize the same discrepancy as before. We accept the functional framework of [8] and thus the resulting optimal control problem is of the type studied in Sec. 4. Theorem 4.1. assures the existence of the solution.

Acknowledgements: The present research work was supported by a research grant in the framework of the Scientific Cooperation between Greece and F.R.G.

REFERENCES

[1] Panagiotopoulos,P.D., Stavroulakis, G. - *A Variational Hemivariational Inequality Approach to the Laminated Plate Theory under Subdifferential Boundary Conditions*, Quart.of Appl. Math. XLVI (1988), 409-430.

[2] Stavroulakis, G., Panagiotopoulos, P.D. - *Laminated orthotropic plates under Subdifferential Boundary Conditions. A Variational-Hemivariational Inequality Approach*, ZAMM 68 (1988), 213-224.

[3] Panagiotopoulos, P.D. - *Non-convex superpotentials in the sense of F.H. Clarke and applications*, Mech.Res.Comm. 8 (1981), 335-340.

[4] Panagiotopoulos, P.D. - *Inequality Problems in Mechanics and Applications. Convex and Non-convex Energy Functions*, Birkhäuser Verlag, Basel, Boston, Stuttgart 1985 (Russian Transl. MIR Publ. Moscow 1988).

[5] Moreau, J.J., Panagiotopoulos, P.D., Strang, G. - *Topics in Nonsmooth Mechanics*, Birkhäuser Verlag, Boston, Basel 1988.

[6] Moreau, J.J., Panagiotopoulos, P.D. (eds.) - *Nonsmooth Mechanics and Applications*, CISM Lect.Notes Vol.302, Springer Verlag, Wien, N.York 1988.

[7] Moreau, J.J. - *La notion de sur-potentiel et les liaisons unilatérales en élastostatique*, C.R.Acad.Sc.Paris 267A (1968), 954-957.

[8] Panagiotopoulos, P.D. - *Variational Hemivariational Inequalities in Nonlinear Elasticity. The convex case*, Aplikace Matem. 33 (1988), 249-268.

[9] Yvon, J.P. - *Etude de quelques problèmes de contrôle pour des systemes distribués*, Thèse de Doctorat D'Etat, Université Paris VI, 1973.

[10] Lions, J.L. - *Optimal control of systems governed by partial differential equations*, Springer-Verlag 1971.

[11] Panagiotopoulos, P.D. - *Optimal control in the Unilateral Thin Plate Theory*, Archives of Mechanics 29 (1977), 25-39.

[12] Mignot, F. - *Contrôle dans les inéquations variationnelles elliptiques*, J.Funct.Anal., 22 (1976) 130-185.

[13] Mignot, F., Puel, J.P. - *Optimal control in some variational inequalities*, SIAM Journal on Control and Optimization 22 (1984), 466-476.

[14] Shuzhong Shi - *Optimal control of Strongly Monotone Variational Inequalities*, SIAM Journal on Control and Optimization 26 (1988), 274-290.

[15] Haslinger, J., Neittaanmäki, P., Tiihonen, T. - *Shape Optimization on Contact Problems Based on Penalization of the State Inequality*. Aplikace Matematiky 31 (1986), 1-88.

[16] Haslinger, J., Neittaanmäki, P. - *On Optimal Shape Design of Systems Governed by Mixed Dirichlet-Signorini Boundary Value Problems*, Math.Meth. in the Appl.Sci. 8 (1986), 157-181.

[17] Haslinger, J., Panagiotopoulos, P.D. - *Optimal Control of Hemivariational Inequalities*, In: Control of Boundaries and Stabilization ed. by J. Simon, Springer-Verlag, N.York Lect.Notes in Control and Information Sciences, Vol. 125, 1989.

[18] Panagiotopoulos, P.D., Haslinger, J. - *Optimal Control of Systems governed by Hemivariational Inequalities*, In: Math. Models for Phase Change Problems ed. by J.F. Rodrigues, Birkhäuser Verlag, Basel, Boston ISNM, Vol. 88, 1989.

[19] Clarke, F.H. - *Nonsmooth analysis and optimization*, J.Wiley, New York 1984.

[20] Chang, K.C. - *Variational methods for non-differentiable functionals and their applications to partial differential equations*, J.Math.Anal.Appl. 80 (1981), 102-129.

[21] Rauch, J. - *Discontinuous semilinear differential equations and multiple valued maps*, Proc. A.M.S. 64 (1977), 277-282.

address:

P.D. Panagiotopoulos

Dept. of Civil Engineering, Aristotle University,

GR-54006 Thessaloniki and

Faculty of Mathematics and Physics, RWTH, D-5100 Aachen

THEOREMS OF THE ALTERNATIVE IN UNILATERAL STRUCTURAL MECHANICS

Giovanni Romano, Luciano Rosati, Giuseppe Ferro

Istituto di Scienza delle Costruzioni, Faculty of Engineering, University of Naples, Italy

1. Introduction

This paper deals with the formulation of some general results in Unilateral Structural Mechanics which can be posed in terms of *alternatives*.

An *alternative* consists of two statements (a) and (b) such that if one of the two statements holds true the other one is false and viceversa. Theorems of the Alternative are well known in Mathematics and are strictly related to existence results. We are interested in suitable extensions of some Theorems of the Alternative in Mathemathical Programming to the context of Structural Analysis.

The paper provides a preliminary short presentation of some classical results in Convex Analysis concerning polarity correspondences between closed convex cones and of an intersection property of closed convex sets recently investigated by the authors [1], [2]. The latter yields necessary and sufficient conditions in order the intersection of two closed convex sets to consist only of points which belong to the relative boundary of at least one of them.

In view of their applications to structural problems with unilateral constraints, these conditions are then specialized to the case in which the two sets are cones. On the basis of the polarity properties of closed convex cones, generalized versions of Farkas' lemma and of Gale's theorem are derived and applied to the discussion of equilibrium and compatibility conditions in unilateral structural problems. The intersection properties of closed convex cones are in turn applied to get generalized formulations of Gordan's and Stiemke's Theorems of the Alternative. These theorems are shown to provide variational conditions of existence for irreversible fields in structural problems with unilateral constraints.

Really surprisingly, suitably revisited and generalized versions of such aged results as Farkas' lemma and Gordan's and Stiemke's Theorems of the Alternative, proved in 1902, 1873 and 1915 respectively, are able to provide an answer to some new problems in structural mechanics involving external and internal unilateral constraints.

2. Preliminary definitions and properties

Let X, X^* be a pair of locally convex linear topological spaces placed in separating duality by a real-valued bilinear form $\langle \cdot, \cdot \rangle$.
A convex set K in X is a subset which is closed with respect to convex combinations of its elements. If K_1 and K_2 are convex sets such is their sum, defined by:

$$K_1 + K_2 = \{x = x_1 + x_2\,;\; x_1 \in K_1,\; x_2 \in K_2\}.$$

The opposite of a convex set K is defined by $-K = \{x \in X \;:\; -x \in K\}$ and we shall write $K_1 - K_2$ as an abbreviated form for $K_1 + (-K_2)$.
A convex cone C in X is a convex set which is closed with respect to the multiplication by a non negative scalar.
The positive polar of C is the closed convex cone in X^* defined by:

$$C^+ = \{x^* \in X^* \mid \langle x^*, x \rangle \geq 0 \quad \forall x \in C\},$$

and the negative polar C^- is the opposite cone of C^+.
Notice that, if C is closed, then $C^{++} = C$. If C is a subspace the positive and negative polars coincide with its orthogonal complement, that is: $C^+ = C^- = C^\perp$.
If K is a closed convex set in X the smallest subspace including K is the linear hull $Lin\,K$ and the largest subspace included in K is the lineality space $lin\,K$.
The outward normal cone to K at a point $x \in K$ is defined by:

$$N_K(x) = \{x^* \in X^* \mid \langle x^*, y - x \rangle \leq 0 \quad \forall y \in K\},$$

and the inward normal cone is defined as the opposite of $N_K(x)$. If C is a closed convex cone in X the lineality space of C is given by: $lin\,C = (-C) \cap C$. It is easy to show that $(Lin\,C)^\perp = C^+ \cap C^- = lin\,C^+ = lin\,C^-$.
The outward normal cone to C at a point $x \in C$ is given by:

$$N_C(x) = \{x^* \in C^- \mid \langle x^*, x \rangle = 0\}.$$

Hence the normal cone to C at the origin is C^-.
Let f be a convex functional defined on a convex set $A = \operatorname{dom} f$ in X. Assuming that f takes the value $+\infty$ outside A, the Fenchel conjugate of f is defined as [3,4,5]:

$$f^*(x^*) = \sup_x \{\langle x^*, x \rangle - f(x)\}.$$

Let A^* be the domain of f^* in X^*. A pair x, x^* is said to be conjugate with respect to the pair f, f^* if Fenchel's equality holds:

$$f(x) + f^*(x^*) = \langle x^*, x \rangle \qquad x \in A,\ x^* \in A^*,$$

or equivalently:

$$\begin{cases} x^* \in \partial f(x), \\ x \in \partial f^*(x^*), \end{cases}$$

where the symbol ∂ denotes the convex differential operator defined by:

$$x^* \in \partial f(x) \overset{\text{def}}{=} f(z) - f(x) \geq \langle x^*, z - x \rangle \quad \forall\, z \in X.$$

2.1 Polarity properties of convex cones

Let X, X^* and Y, Y^* be two pairs of spaces in duality and T, T^* be a dual pair of linear continuous operators:

$$T : X \mapsto Y,$$

$$T^* : Y^* \mapsto X^*,$$

with:

$$\langle y^*, Tx \rangle = \langle T^* y^*, x \rangle \quad \forall x \in X, y^* \in Y^*.$$

By $\mathcal{N}(T)$ and $\mathcal{R}(T)$ we shall denote the null space and the range space of T respectively.
If C_1 and C_2 are convex cones in X we have:

$$(C_1 + C_2)^+ = C_1^+ \cap C_2^+. \tag{1}$$

Moreover if C_X and C_Y are convex cones in X and in Y respectively, then

$$(TC_X)^+ = T^{*-1} C_X^+, \tag{2}$$

$$(T^* C_Y^+)^+ = T^{-1} C_Y. \tag{3}$$

Useful relations between polar convex cones are provided by the following results:

Lemma 1. *If the convex cones C_1 and C_2 are closed we have:*

$$C_1^+ + C_2^+ = (C_1 \cap C_2)^+. \tag{4}$$

Lemma 2. *If the convex cones TC_X and $T^*C_Y^+$ are closed, we have:*

$$TC_X = (T^{*-1}C_X^+)^+, \tag{5}$$

$$T^*C_Y^+ = (T^{-1}C_Y)^+. \tag{6}$$

According to a theorem due to Fenchel [3,4,5], K closed implies TK closed if the following sufficient (but unfortunately not necessary) condition is met:

$$\mathcal{N}(T) \cap K_\infty \subset lin\, K,$$

where K_∞ is the recession cone of K, that is the closed convex cone of the elements $y \in X$ such that $x + \lambda y \in K$ for every $\lambda \geq 0$ and $x \in K$. Its elements are also called directions of recession of K and it is possible to prove that:

$$K_\infty = \{\, y \in X \;:\; K + y \subset K \}.$$

A relevant special case in which K closed implies TK closed is that of K polyhedral [5].
In Mathemathical Programming K is the closed positive orthant. Since this cone is polyhedral, TK is closed for any linear operator T.

2.2 Intersection properties of convex cones

If the linear hull $Lin\, C$ is strictly included in the ambient space X, every point of C belongs to the boundary *bnd* C of C. In this case it is convenient to introduce the concept of relative boundary $relbnd\, C$, which is the boundary of C when $Lin\, C$ is considered as the ambient space.
Alternatively the points of the relative boundary of C can be defined as those characterized by the existence of a supporting hyperplane which does not include C.
In formulae:

$$x_o \in relbnd\, C \iff \exists\, x_o^* \in N_C(x_o), \quad x_o^* \notin (Lin\, C)^\perp. \tag{7}$$

The relative interior $relint\; C$ of C is the subset of the points in C which do not belong to the relative boundary of C.
If A and B are closed convex sets with non empty intersection in X, the intersection of the outward normal cone to A and of the inward normal cone to B at each point of $A \cap B$ results to be equal to the normal cone $N_{A-B}(0)$.
That is:

$$A \cap B \neq \emptyset \implies N_{A-B}(0) = N_A(z_o) \cap -N_B(z_o) \quad \forall z_o \in A \cap B. \tag{8}$$

This property is the basis of the following result [1,2]:

Lemma 3. *The non empty intersection of two closed convex sets A and B consists of points which belong to the boundary of both sets and to the relative boundary of at least one of them if and only if:*

$$\exists\, z_o^* \in [N_A(z_o) \cap -N_B(z_o)] \cup [N_B(z_o) \cap -N_A(z_o)] \quad \forall z_o \in A \cap B, \tag{9}$$

with z_o^* such that at least one of the following conditions is met:

$$\begin{cases} z_o^* \notin (Lin\, A)^{\perp}, \\ z_o^* \notin (Lin\, B)^{\perp}, \end{cases} \tag{10}$$

or equivalently $z_o^* \notin (Lin\, A)^{\perp} \cap (Lin\, B)^{\perp}$.

The above conditions admit the following geometrical interpretation: *it exists a supporting hyperplane separating A and B which does not include both sets.*
If A and B are closed convex cones, trivially $0 \in A \cap B$ so that:

$$N_{A-B}(0) = N_A(0) \cap -N_B(0) = A^- \cap B^+$$

and the previous lemma can be specialized into the following one.

Lemma 4. *The intersection of two closed convex cones A and B consists of points which belong to the boundary of both sets and to the relative boundary of at least one of them if and only if:*

$$\exists\, z_o^* \in A^- \cap B^+ \tag{11}$$

with z_o^* such that at least one of the following conditions is met:

$$\begin{cases} z_o^* \notin A^+ \cap A^-, \\ z_o^* \notin B^+ \cap B^-. \end{cases} \tag{12}$$

Depending on whether the first, the second or both conditions (10), (12) are satisfied, the following intersection properties will respectively hold true:

$$A \cap B = relbnd\, A \cap bnd\, B,$$
$$A \cap B = bnd\, A \cap relbnd\, B,$$
$$A \cap B = relbnd\, A \cap relbnd\, B,$$

or, equivalently :

$$relint\, A \cap bnd\, B = \emptyset,$$
$$bnd\, A \cap relint\, B = \emptyset,$$
$$relint\, A \cap relint\, B = \emptyset.$$

3. Classical Theorems of the Alternative

The Theorems of the Alternative in Mathematical Programming are usually formulated in a purely algebraic framework in terms of vectors and matrices. A comprehensive treatment and an exhaustive list of this kind of results can be found e.g. in [6].
To present their classical formulations let us consider an $m \times n$ matrix A:

$$A : \mathcal{R}^n \mapsto \mathcal{R}^m$$

and its transpose A^*. Let us take as C the positive orthant in $\mathcal{R}^n$:

$$C = \{x \in \mathcal{R}^n : x_i \geq 0\},$$

where x_i denotes the ith component of the vector x.
The positive orthant C is a very special cone since it is polyhedral and its space of lineality $lin\, C = (-C) \cap C$ degenerates into the null vector.
The following notations will be adopted:

$$x \geqq 0 \iff x_i \geq 0 \quad \forall i,$$
$$x \geq 0 \iff x \geqq 0 \quad \text{and} \quad x \neq 0,$$
$$x > 0 \iff x_i > 0 \quad \forall i.$$

The vector x is said to be non negative if $x \geqq 0$ ($x \in C$); it is said to be semipositive if $x \geq 0$ [$x \in C$; $x \notin (-C) \cap C$], and is said to be positive if $x > 0$ ($x \in int\ C$).
Farkas' lemma consists of the following alternative:

either

$$a) \quad Ax = b, \quad x \geqq 0 \text{ has a solution } x \in \mathcal{R}^n$$

or

$$b) \quad A^*y \geqq 0, \quad \langle b, y \rangle < 0 \text{ has a solution } y \in \mathcal{R}^m.$$

A more general theorem by **Gale** yields the following alternative:

either

$$a) \quad Ax \leqq b, \quad x \geqq 0 \text{ has a solution } x \in \mathcal{R}^n$$

or

$$b) \quad A^*y \geqq 0, \quad y \geqq 0, \quad \langle b, y \rangle < 0 \text{ has a solution } y \in \mathcal{R}^m.$$

Different kinds of alternatives are provided by the following theorems:

Gordan's theorem:

either

$$a) \quad Ax = 0, \quad x \geq 0 \text{ has a solution } x \in \mathcal{R}^n$$

or

$$b) \quad A^* y > 0 \text{ has a solution } y \in \mathcal{R}^m.$$

Stiemke's theorem:

either

$$a) \quad Ax = 0, \quad x > 0 \text{ has a solution } x \in \mathcal{R}^n$$

or

$$b) \quad A^* y \geq 0 \text{ has a solution } y \in \mathcal{R}^m.$$

It has to be remarked that, when the positive orthant is replaced by a subspace, **Farkas'** lemma and **Gale's** theorem can be reformulated in a simpler special form while **Gordan's** and **Stiemke's** theorems cannot.

In the next section these results will be formulated and proved in a more general context by means of a geometrical approach.

4. Generalized Theorems of the Alternative

The generalization of the Theorems of the Alternative presented above is performed in two directions. First, we deal with closed convex cones other than the positive orthant of $\mathcal{R}^n$ and, second, the ambient spaces will be allowed to be infinite dimensional locally convex topological vector spaces.

In this more general framework a geometrical approach instead of a numerical one will be followed, both in the statement and in the proof of the theorems. In the authors' opinion, this approach allows for a more synthetic and intuitive exposition of the results.

The generalized **Farkas'** lemma and **Gale's** theorem follow directly from the polarity properties proved in lemma 1 and lemma 2. Reference is made to section 2 for notations and definitions.

Generalized Farkas' lemma. *Let C_X be a closed convex cone in X and let y_o be an assigned vector in Y.*
If TC_X is a closed convex cone, the following alternative holds:

either

$$a) \quad \exists\, x \in C_X : \quad Tx = y_o,$$

or

$$b) \quad \exists\, y^* \in Y^* : \quad T^* y^* \in C_X^+; \quad \langle y^*, y_o \rangle < 0.$$

Proof. *We have to show that: b) $false \iff$ a) $true$. Now:*

$$b)\, false \iff \langle y^*, y_o \rangle \geq 0 \quad \forall y^* \in Y^* : \quad T^* y^* \in C_X^+ \iff y_o \in (T^{*-1} C_X^+)^+$$

and

$$a)\, true \iff y_o \in TC_X.$$

The alternative is then equivalent to the statement:

$$(T^{*-1} C_X^+)^+ = TC_X,$$

which is the equality (5) in Lemma 2. □

Generalized Gale's theorem. *Let C_X and C_Y be closed convex cones in X and Y and y_o an assigned vector in Y. If TC_X is a closed convex cone, the following alternative holds:*

either

$$a) \quad \exists\, x \in C_X : \quad Tx \in y_o - C_Y,$$

or

$$b) \quad \exists\, y^* \in C_Y^+ : \quad T^* y^* \in C_X^+; \quad \langle y^*, y_o \rangle < 0.$$

Proof. *We have to show that: b) $false \iff$ a) $true$. Now:*

$$b)\, false \iff \langle y^*, y_o \rangle \geq 0 \quad \forall y^* \in C_Y^+ : \quad T^* y^* \in C_X^+ \iff y_o \in (T^{*-1} C_X^+ \cap C_Y^+)^+$$

and

$$a)\, true \iff y_o \in TC_X + C_Y.$$

The alternative is then equivalent to the statement:

$$(T^{*-1} C_X^+ \cap C_Y^+)^+ = TC_X + C_Y,$$

which is an immediate consequence of Lemma 1 and Lemma 2, equality (5). □

Generalized versions of **Gordan's** and **Stiemke's** theorems can be proved on the basis of the intersection properties discussed in lemma 4.

Generalized Gordan's theorem. *If C_X is a closed convex cone in X then:*

either

$$a) \quad \exists\, x \in C_X \cap \mathcal{N}(T): \quad x \notin (Lin\, C_X^+)^\perp = (-C_X) \cap C_X,$$

or

$$b) \quad \mathcal{R}(T^*) \cap relint\ C_X^+ \neq \emptyset.$$

Proof. *It is sufficient to remark that, by Lemma 4, statement a) is equivalent to:*

$$\mathcal{R}(T^*) \cap relint\ C_X^+ = \emptyset.$$

□

Generalized Stiemke's theorem. *If C_X is a closed convex cone in X then:*

either

$$a) \quad \exists\, x^* \in C_X^+ \cap \mathcal{R}(T^*): \quad x^* \notin (Lin\, C_X)^\perp = C_X^+ \cap C_X^-,$$

or

$$b) \quad \mathcal{N}(T) \cap relint\ C_X \neq \emptyset.$$

Proof. *It is sufficient to remark that, by Lemma 4, statement a) is equivalent to:*

$$\mathcal{N}(T) \cap relint\ C_X = \emptyset.$$

□

It is interesting to note that obvious variants of **Farkas'** lemma and of **Gale's** theorem can be deduced from equality (6) in Lemma 2. In the same way, obvious variants of **Gordan's** and **Stiemke's** theorems can be deduced by discussing the intersections $\mathcal{R}(T) \cap relint\ C_Y$ and $\mathcal{N}(T^*) \cap relint\ C_Y^+$ respectively.

It is moreover evident that the positive polars referred to in the formulation of the theorems can be changed into the negative ones.

We remark that the relative interior of a closed convex cone C_X can be characterized as follows:

$$\begin{aligned} relint\ C_X &= \{x \in C_X : \quad N_{C_X}(x) = (Lin\, C_X)^\perp\} = \\ &= \{x \in C_X : \nexists\, x^* \notin C_X^+, x^* \notin (Lin\, C_X)^\perp, \langle x^*, x\rangle = 0\} = \\ &= \{x \in C_X : \quad \langle x^*, x\rangle > 0 \quad \forall x^* \in C_X^+, x^* \notin C_X^+ \cap C_X^-\}. \end{aligned}$$

In finite dimension C_X is the closed positive orthant of $\mathcal{R}^n$, so that $C_X^+ = C_X$ and $(Lin\, C_X)^\perp = C_X^+ \cap C_X^-$ is the null vector. Hence:

$$x \in relint\ C_X \iff x > 0,$$

and:

$$x^* \notin C_X^+ \cap C_X^- \iff x^* \neq 0,$$

so that the statements in the theorems above do reduce to their classical formulations.

5. Theorems of the Alternative in Unilateral Structural Mechanics

The generalized theorems of the alternative developed in the previous paragraph can be applied to provide existence conditions for equilibrium and compatibility problems in unilateral structural mechanics.

Unilateral structural problems are characterized by admissible sets for the state variables which are closed convex conical manifolds. By definition, a conical manifold consists of the translation of a cone.

The analysis is developed in the context of a geometrically linearized theory so that the equilibrium and the compatibility operators are dual linear operators.

A unilateral problem for the structural model is defined by the following items:

- **State variables and ambient spaces**

$$\begin{cases} \left.\begin{array}{ll} u \in \mathcal{V} & \text{displacements} \\ f \in \mathcal{F} & \text{external forces} \end{array}\right\} \text{external variables} \\ \left.\begin{array}{ll} \varepsilon \in \mathcal{D} & \text{deformations} \\ \sigma \in \mathcal{S} & \text{internal forces} \end{array}\right\} \text{internal variables} \end{cases} \tag{13}$$

where $\begin{cases} \mathcal{V}, \mathcal{F} \\ \mathcal{D}, \mathcal{S} \end{cases}$ are dual pairs of linear spaces.

- **Field equations**

$$\begin{cases} Tu = \varepsilon & \text{kinematic compatibility} \\ T^*\sigma = f & \text{static equilibrium} \end{cases} \tag{14}$$

where :

$$\begin{cases} T : \mathcal{V} \to \mathcal{D} \\ T^* : \mathcal{S} \to \mathcal{F} \end{cases} \tag{15}$$

is a dual pair of linear operators.

- **Constraints**

$$\begin{cases} \left.\begin{array}{l} f \in \partial \jmath(u) \\ u \in \partial \jmath^*(f) \end{array}\right\} \text{external constraints} \\ \left.\begin{array}{l} \sigma \in \partial \varphi(\varepsilon) \\ \varepsilon \in \partial \varphi^*(\sigma) \end{array}\right\} \text{internal constraints} \end{cases} \tag{16}$$

where $\jmath(u), \jmath^*(f)$ is a conjugate pair of concave external potentials and $\varphi(\varepsilon), \varphi^*(\sigma)$ is a conjugate pair of convex internal potentials.

- **Admissible sets**

$$\begin{array}{ll} U = \operatorname{dom} \jmath & \text{admissible displacements} \\ P = \operatorname{dom} \jmath^* & \text{admissible external forces} \\ E = \operatorname{dom} \varphi & \text{admissible deformations} \\ Q = \operatorname{dom} \varphi^* & \text{admissible internal forces} \end{array} \tag{17}$$

where $U \subset \mathcal{V}, P \subset \mathcal{F}, E \subset \mathcal{D}, Q \subset \mathcal{S}$ are closed convex sets.
In the present unilateral framework the potentials $\jmath, \jmath^*, \varphi, \varphi^*$ are assumed to be such that the admissible sets U, P, E, Q are closed convex conical manifolds. A discussion on the conditions for the static and kinematic admissibility of structural problems with convex constraints and for the attainment of the corresponding limit states can be found in [1,2]. The results presented in this paper are the specialization of those conditions to unilateral structural problems.

5.1 Static equilibrium and kinematic compatibility

Farkas' lemma can be directly applied to get either the static equilibrium condition in structural problems with unilateral constraints on the internal forces or the kinematic compatibility condition in structural problems with unilateral constraints on the displacements.

Static equilibrium condition. *If Q is a closed convex cone and if the cone T^*Q is closed, then for any given external force $f \in \mathcal{F}$ the following alternative holds:*

either

$$a) \quad \exists\, \sigma \in Q : \quad T^*\sigma = f,$$

or (18)

$$b) \quad \exists\, u \in \mathcal{V} : \quad Tu \in Q^-; \quad \langle f, u \rangle > 0.$$

Kinematic compatibility condition. *If U is a closed convex cone and if the cone TU is closed, then for any given deformation $\varepsilon \in \mathcal{D}$ the following alternative holds:*

either

$$a) \quad \exists\, u \in U : \quad Tu = \varepsilon,$$

or (19)

$$b) \quad \exists\, \sigma \in \mathcal{S} : \quad T^{*}\sigma \in U^{+}; \quad \langle \sigma, \varepsilon \rangle < 0.$$

The alternatives (18) and (19) are equivalent to the polarity relations respectively:

$$T^{*}Q = (T^{-1}Q^{-})^{-} \tag{20}$$

$$TU = (T^{*-1}U^{+})^{+}. \tag{21}$$

The theorem by **Gale** yields a more general equilibrium and compatibility condition for structural problems with unilateral constraints on the external and internal forces and on the deformations and displacements.

Static equilibrium condition. *Let the admissible set for the external forces be the conical manifold $P = \ell + R$ where ℓ is a given load and R is the closed convex cone of all admissible reactions. If Q is a closed convex cone and if the cone $T^{*}Q$ is closed, the following alternative holds:*

either

$$a) \quad \exists\, \sigma \in Q : \quad T^{*}\sigma \in \ell + R,$$

or (22)

$$b) \quad \exists\, u \in R^{+} : \quad Tu \in Q^{-}; \quad \langle \ell, u \rangle > 0.$$

Kinematic compatibility condition. *Let the admissible set for the deformations be the conical manifold $E = \delta + \Delta$ where δ is an imposed distorsion and Δ is the closed convex cone of admissible additional deformations. If U is a closed convex cone and if the cone TU is closed, the following alternative holds:*

either

$$a) \quad \exists\, u \in U : \quad Tu \in \delta + \Delta,$$

or (23)

$$b) \quad \exists\, \sigma \in \Delta^{-} : \quad T^{*}\sigma \in U^{+}; \quad \langle \sigma, \delta \rangle < 0.$$

The alternatives (22) and (23) are equivalent to the polarity relations:

$$T^*Q - R = (T^{-1}Q^- \cap R^+)^-, \tag{24}$$

$$TU - \Delta = (T^{*-1}U^+ \cap \Delta^-)^+, \tag{25}$$

respectively.

5.2 Existence conditions for irreversible fields

The theorems by **Gordan** and by **Stiemke** can be invoked to provide existence conditions for irreversible fields.
Assuming that the admissible sets U, P, E, Q are closed convex cones, we define as *irreversible* those fields which do not belong to the lineality space of the corresponding admissible sets.
The **Gordan's** theorem supplies a necessary and sufficient condition for the existence of an irreversible rigid displacement or of an irreversible selfequilibrated internal force.

Existence of an irreversible rigid displacement.

either

$$a) \quad \exists\, u \in U \cap \mathcal{N}(T), \quad u \notin -U \cap U$$

or (26)

$$b) \quad \mathcal{R}(T^*) \cap relint\, U^+ \neq \emptyset.$$

Existence of an irreversible selfequilibrated internal force.

either

$$a) \quad \exists\, \sigma \in Q \cap \mathcal{N}(T^*), \quad \sigma \notin -Q \cap Q$$

or (27)

$$b) \quad \mathcal{R}(T) \cap relint\, Q^- \neq \emptyset.$$

Similarly, the **Stiemke's** theorem yields a necessary and sufficient condition for the existence of an an irreversible kinematically compatible deformation or of an irreversible equilibrated external force.

Existence of an irreversible kinematically compatible deformation.

either

$$a) \quad \exists\, \varepsilon \in E \cap \mathcal{R}(T), \quad \varepsilon \notin -E \cap E$$

or (28)

$$b) \quad \mathcal{N}(T^{*}) \cap relint\ E^{-} \neq \emptyset.$$

Existence of an irreversible equilibrated external force.

either

$$a) \quad \exists\, f \in P \cap \mathcal{R}(T^{*}), \quad f \notin -P \cap P$$

or (29)

$$b) \quad \mathcal{N}(T) \cap relint\ P^{+} \neq \emptyset.$$

The mechanical interpretation of the b) alternatives in the previous theorems (26),(27),(28) and (29) can be inferred from the characterization of the relative interior of a convex set reported at the end of section 4.

6. Applications

Interesting applications of the theorems above can be found in the mechanics of materials which are subject to unilateral conditions on the stress fields.
Relevant examples are provided by the *no tension* materials and by cohesionless materials with a frictional behaviour obeying to Coulomb's limit condition. In both cases, the admissible set for the internal forces is a closed convex cone whose linear hull is the whole ambient space and with a lineality space degenerating into the null vector. The **Gordan's** theorem (27) then states that either a non trivial admissible and selfequilibrated internal force does exist or there is a kinematically compatible deformation for which every non trivial admissible internal force performs a negative virtual work.
Such a deformation field can be easily found to be given by an isotropic expansion and hence we may state the following:

Nonexistence theorem. *In a structure made of no tension material or of a cohesionless material with a frictional behaviour obeying Coulomb's limit condition, the existence of a nontrivial selfequilibrated admissible internal force is ruled out.* □

Similar results can be inferred by the other theorems of the alternative and can be applied to the discussion of unilateral problems in structural mechanics.

Acknowledgements. *This research has been developed with the financial support of the M.P.I. of Italy.*

REFERENCES

1. G. Romano, L. Rosati , On Limit states, Plasticity and Locking. *Atti IX Congresso Naz. AIMETA*, Bari, October 4-7, 1988.
2. L. Rosati , Problems of Structural Mechanics with Convex Constraints. *Doctoral Dissertation*, Naples, 1989.
3. W. Fenchel, On conjugate convex functions. *Canad. J. Math.* , **1**, pag. 73-77, 1949.
4. W. Fenchel, Convex Cones, Sets and Functions. *Lecture Notes*, Princeton University, 1951.
5. R. T. Rockafellar, *Convex Analysis*. Princeton University Press, Princeton, 1970.
6. O. L. Mangasarian, *Nonlinear Programming*. Mc Graw Hill, New York, 1969.

Giovanni Romano
Istituto di Scienza delle Costruzioni,
Faculty of Engineering, University of Naples,
Piazzale Tecchio, 80122 Naples, Italy
Phone : 39 - 81 - 7682110
Telex : 39 - 81 - 722392 - INGENA I
Telefax : 39 - 81 - 610445

International Series of Numerical Mathematics, Vol. 101, © 1991 Birkhäuser Verlag Basel

QUASI-STATIC SIGNORINI'S CONTACT PROBLEM WITH FRICTION AND DUALITY

Józef Joachim Telega

Polish Academy of Sciences, Institute of Fundamental Technological Research, Warsaw, Poland

Introduction

Contact problems with friction cannot be formulated, at the present stage of our knowledge, as extremum problems. Hence usual duality approaches, as presented for instance in [4,8,14] do not apply to such problems. Capuzzo-Dolcetta and Matzeu [3] proposed the theory of duality for so called implicit variational problems. For the sake of simplicity we call it M-CD-M duality theory, see [16] for more details.

In the static case, the variational formulation of the Signorini problem with friction is available in the form of one implicit variational inequality [5,15,16]. The M-CD-M duality theory was applied to such a unilateral contact problem in the paper [2], see also [15]. Detailed study of duality is contained in the paper [16].

Klarbring, Mikelić and Shillor [6,7] studied duality for frictional contact problems with normal compliance. The same theory of duality was used.

Evolution (quasi-static and dynamic) problems with friction are more complicated because their variational formulations are not available in the form of a single inequality. However, as suggested in [16], the variational formulation of such problems is readily obtained in the form of an implicit variational inequality coupled with a variational inequality. The present contribution deals with duality for the quasi-static Signorini problem with friction. To achieve this aim the M-CD-M duality theory has first been extended to a system consisting of two coupled inequalities, completed by an initial condition.

1. Strong and variational formulations of the quasi-static Signorini's problem with friction

Let $\Omega \subset R^3$ be a sufficiently regular domain and $\Gamma = \partial\Omega$ its boundary. The latter consists of three nonoverlapping parts: Γ_0, Γ_1 and Γ_2, such that $\Gamma = \overline{\Gamma_0} \cup \overline{\Gamma_1} \cup \overline{\Gamma_2}$ and the surface measure of Γ_0 is positive. Here the bar over a set denotes its closure. Γ_2 is the surface of a possible contact with a rigid foundation.

Let $\mathbf{n} = (n_i)$ denote a unit exterior normal vector to Γ. Latin indices run from 1 to 3. Two-dimensional problems can be treated similarly. A vector $\mathbf{v} = (v_i)$ defined on a part of Γ may be decomposed as follows:

$$\mathbf{v} = v_n\mathbf{n} + \mathbf{v}_t, \tag{1.1}$$

where $v_n = v_i n_i$ denotes the normal component of $\mathbf{v}$, while $v_{ti} = v_i - v_n n_i$ are its tangential components.

A similar decomposition can be performed for a stress vector $(s_{ij}n_j)$, where $\mathrm{s} = (s_{ij})$ is a stress tensor. Thus we have

$$s_{ij}n_j = s_n n_i + s_{ti}, \tag{1.2}$$

where $s_n = s_{ij}n_i n_j$ and $s_{ti} = s_{ij}n_j - s_n n_i$.
Let K be a closed set of frictionally admissible normal and tangential stresses on Γ_2. Usually, this set is given by

$$K = \{\boldsymbol{\sigma} = (\sigma_n, \boldsymbol{\sigma}_t) \mid g_J(\sigma_n, \boldsymbol{\sigma}_t) \leq 0, \quad J = 1, ..., M\}. \tag{1.3}$$

The functions g_J may be anisotropic and nonconvex. By $K(s_n)$ we denote a section of K with a plane $s_n =$ consts. For instance, if K is defined by (1.3), then we have

$$K(s_n) = \{\mathbf{t} \mid g_J(s_n, \mathbf{t}) \leq 0, \quad \mathbf{t}\cdot\mathbf{n} = 0, \quad J = 1, ..., M\}. \tag{1.4}$$

We make the following assumption:

(A) | for a fixed s_n the closed set $K(s_n)$ is convex and bounded.

$I_{K(s_n)}$ denotes the indicator function of the set $K(s_n)$, i.e.:

$$I_{K(s_n)}(\mathbf{t}) = \begin{cases} 0, & \text{if } \mathbf{t} \in K(s_n), \\ +\infty, & \text{otherwise.} \end{cases} \tag{1.5}$$

The friction law or sliding rule is assumed in the following subdifferential form [15,16]

$$-\dot{\mathbf{u}}_t \in \partial I_{K(s_n)}(\mathbf{s}_t), \tag{1.6}$$

where the dot denotes differentiation with respect to time and $\partial I_{K(s_n)}(\mathbf{s}_t)$ is the subdifferential of the indicator function $I_{K(s_n)}$ at $\mathbf{s}_t$. The subdifferential law (1.6) can equivalently be written in the following form

$$(\mathbf{t} - \mathbf{s}_t)\cdot\dot{\mathbf{u}}_t \geq 0, \quad \forall\, \mathbf{t} \in K(s_n), \tag{1.7}$$

and

$$\mathbf{s}_t \in \partial_2 d(s_n, -\dot{\mathbf{u}}_t). \tag{1.8}$$

Here $d(s_n, .)$ is the support function of the set $K(s_n)$. It is given by [13]

$$d(s_n, \mathbf{v}) = \sup\{\mathbf{v}\cdot\mathbf{t} \mid \mathbf{t} \in K(s_n)\}. \tag{1.9}$$

Moreover, $\partial_2 d(s_n, .)$ is the subdifferential of the convex and lower semicontinuous (in fact continuous) function $d(s_n, .)$.

Let us pass to the formulation of the quasi-static Signorini problem with friction. Thus body forces $\mathbf{B}$ and tractions $\mathbf{F}$ depend also on time $\tau \in [0, \tau_0]$, $\tau_0 > 0$, i.e.: $\mathbf{B} = \mathbf{B}(x, \tau)$, $\mathbf{F} = \mathbf{F}(x, \tau)$ where x is a space variable. Here we consider only quasi-static loadings, though a generalization of our considerations to the dynamic case is possible. Usually we shall write $\mathbf{u}(\tau)$ instead of $\mathbf{u}(x, \tau)$; in other words we set $\mathbf{u}(\tau) = \{\mathbf{u}(x, \tau),\ x \in \Omega\}$, see [9]. The strong formulation of the initial-boundary value problem considered has the form of the Problem P_1

Find $\mathbf{u}(\tau)$ $(\tau \in [0, \tau_0])$ such that

$$s_{ij,j}(\mathbf{u}(\tau)) + B_i(\tau) = 0, \quad \text{in } \Omega \times [0, \tau_0], \tag{1.10}$$

$$s_{ij}(\mathbf{u}(\tau)) = a_{ijkl} e_{kl}(\mathbf{u}(\tau)), \tag{1.11}$$

$$\mathbf{u}(\tau) = 0, \quad \text{on } \Gamma_0 \times [0, \tau_0], \tag{1.12}$$

$$s_{ij}(\mathbf{u}(\tau)) n_j = F_i(\tau), \quad \text{on } \Gamma_1 \times [0, \tau_0], \tag{1.13}$$

$$\left.\begin{array}{l} u_n \le 0,\ s_n \le 0,\ s_n u_n = 0 \\ -\dot{\mathbf{u}}_t \in \partial I_{K(s_n)}(\mathbf{s}_t) \end{array}\right\} \text{on } \Gamma_2 \times [0, \tau_0], \tag{1.14}$$

$$\mathbf{u}(0) = 0, \tag{1.15}$$

where

$$e_{ij}(\mathbf{u}) = (u_{i,j} + u_{j,i})/2, \quad u_{i,j} = \frac{\partial u_i}{\partial x_j} \quad .$$

Now we pass to the variational formulation. We make the following assumptions:

$$\exists c_o > 0, \qquad a_{ijkl} e_{ij} e_{kl} \ge c_o e_{ij} e_{ij}, \ \ \forall \mathbf{e} \in E_s^3, \tag{1.16}$$

$$a_{ijkl} = a_{jikl} = a_{klij}, \quad a_{ijkl} \ \text{ is time independent}, \tag{1.17}$$

$$B_i \in L^2[(0, \tau_0), L^2(\Omega)], \quad F_i \in L^2[(0, \tau_0), L^2(\Gamma_1)], \tag{1.18}$$

Here E_s^3 is the space of symmetric 3×3 matrices. It seems not possible to cast the relations (1.10)-(1.14) into one (implicit) variational inequality, because in the Signorini conditions displacement is involved while the sliding rule is of the parabolic type.

We define

$$V = \{\mathbf{v} \in H^1(\Omega, R^3) \mid \mathbf{v} = 0, \text{ on } \Gamma_0\}, \tag{1.19}$$

$$K_n = \{z \mid z \le 0,\ \ z = (\gamma \mathbf{z})_{n|\Gamma_2},\ \ \mathbf{z} \in V\}, \tag{1.20}$$

$$H_s = \{\boldsymbol{\sigma} = (\sigma_{ij}) \mid \sigma_{ij} \in L^2(\Omega),\ \ \sigma_{ij,j} \in L^2(\Omega)\}, \tag{1.21}$$

where γ is the trace operator. The variational formulation, in terms of displacements, has the form of the

Problem P_d

Find $\mathbf{u}(\tau) \in V$ $(\tau \in [0, \tau_0])$ such that $\dot{\mathbf{u}}(\tau) \in V$ and

$$\int_\Omega a_{ijkl} e_{kl}(\mathbf{u}(\tau)) e_{ij}(\mathbf{v} - \dot{\mathbf{u}}(\tau))dx + J(\mathbf{u}(\tau), \mathbf{v}) - J(\mathbf{u}(\tau), \dot{\mathbf{u}}(\tau)) -$$

$$- \int_{\Gamma_2} s_n(\mathbf{u}(\tau))(v_n - \dot{u}_n(\tau))d\Gamma \geq \mathcal{L}(\mathbf{v} - \dot{\mathbf{u}}(\tau)), \ \forall \mathbf{v} \in V, \tag{1.22}$$

$$\int_{\Gamma_2} s_n(\mathbf{u}(\tau))(w - u_n(\tau))d\Gamma \geq 0, \ \ \forall w \in K_n, \tag{1.23}$$

$$\mathbf{u}(0) = 0, \tag{1.24}$$

where

$$J(\mathbf{u}, \mathbf{v}) = \int_{\Gamma_2} d(s_n(\mathbf{u}), -\mathbf{v}_t)d\Gamma, \tag{1.25}$$

$$\mathcal{L}(\mathbf{v}) = \int_\Omega B_i v_i dx + \int_{\Gamma_1} F_i v_i d\Gamma. \tag{1.26}$$

To obtain the implicit variational inequality (1.22) we multiply (1.10) by $(\mathbf{v} - \dot{\mathbf{u}}(\tau))$, integrate over Ω and next proceed in the usual way, see [16]. The variational inequality (1.23) is the global form of the Signorini conditions. We note that $s_n u_n = 0$ and $s_n w \geq 0$ for each $w \in K_n$. It remains to comment the initial condition (1.24), which for the sake of simplicity is assumed as homogeneous. If $\mathbf{u}, \dot{\mathbf{u}} \in L^p[(0, \tau_0), V]$ $(1 \leq p \leq \infty)$ then we have $\mathbf{u} \in C([0, \tau_0], V)$, see [9]. It means that a function $\mathbf{u}$ solving P_d is continuous on the time interval $[0, \tau_0]$. Hence $\mathbf{u}(0)$ makes sense.

Andersson [1] considered existence of a solution for the quasi-static frictional problem with normal compliance. Such a problem is simpler than the frictional Signorini problem investigated here. Andersson obtained his existence theorem in the space $BV[(0, \tau_0), V]$. Such a solution is discontinuous in time. The same will probably be true for the Signorini problem with friction.

We note that in the implicit variational inequality (1.22) $\mathbf{v}$ denotes a virtual velocity field. Further, the variational inequality (1.23) can be written in the following equivalent form

$$\int_{\Gamma_2} s_n(\mathbf{u}(\tau))(w - u_n(\tau))d\Gamma + I_{K_n}(w) - I_{K_n}(u_n(\tau)) \geq 0, \ \ \forall w \in V_1, \tag{1.27}$$

where

$$V_1 = \{w_{|\Gamma_2} \mid w \in H^{1/2}(\Gamma), \ \ w = 0, \ \text{on } \Gamma_0\}. \tag{1.28}$$

We shall now consider the stress approach. The constitutive equation (1.11) and the initial condition (1.15) imply

$$\mathbf{s}(0) = 0. \tag{1.29}$$

We define the set $K_s^0(s_n)$ of admissible stresses:

$$K_s^0(s_n) = \{\boldsymbol{\sigma} \in H_s \mid \sigma_{ij,j} + B_i = 0 \text{ in } \Omega;\ \ \sigma_{ij}n_j = F_i, \text{ on } \Gamma_1;$$
$$\sigma_n = s_n, \boldsymbol{\sigma}_t \in K(s_n), \text{ on } \Gamma_2\}. \tag{1.30}$$

We observe that in (1.30) the condition $\sigma_n \leq 0$ is absent. This condition is taken into account in the inequality (1.36) below. Let $\mathbf{s}(\tau)$ be a stress field and $\mathbf{u}(\tau)$ a displacement field solving the problem P_1. Thus we have

$$0 = \int_\Omega (\sigma_{ij,j} - s_{ij,j})\dot{u}_i dx = -b(\dot{\mathbf{s}}(\tau), \boldsymbol{\sigma} - \mathbf{s}(\tau)) + \int_{\Gamma_2} (\sigma_t - \mathbf{s}_t(\tau)) \cdot \mathbf{u}_t(\tau) d\Gamma,$$

where $\dot{s}_{ij}(\tau) = a_{ijkl}e_{kl}(\dot{\mathbf{u}}(\tau))$; we recall that elastic moduli are time independent. Taking account of the friction law (1.7) we obtain

$$b(\dot{\mathbf{s}}(\tau), \boldsymbol{\sigma} - \mathbf{s}(\tau)) \geq 0,\ \forall \boldsymbol{\sigma} \in K_s^0(s_n), \tag{1.31}$$

where

$$b(\mathbf{s}, \boldsymbol{\sigma}) = \int_\Omega b_{ijkl} s_{ij} \sigma_{kl} dx,\ \ \mathbf{b} = \mathbf{a}^{-1}. \tag{1.32}$$

In terms of stresses the global form of the Signorini conditions is

$$\int_{\Gamma_2} (\tilde{s} - s_n(\tau)) u_n(\mathbf{s}(\tau)) d\Gamma \geq 0,\ \ \forall \tilde{s} \in -K_n^*, \tag{1.33}$$

where

$$K_n^* = \{z^* \in H^{-1/2}(\Gamma_2) \mid < z^*, z >_{|\Gamma_2} \leq 0,\ \forall z \in K_n\}. \tag{1.34}$$

Summarizing we have
<u>Problem Q_s</u>

Find $\mathbf{s}(\tau) \in K_s^0(s_n(\tau))$ $(\tau \in [0, \tau_0])$ such that
$\dot{\mathbf{s}}(\tau) \in L^2(\Omega, E_s^3)$ and

$$b(\dot{\mathbf{s}}(\tau), \boldsymbol{\sigma} - \mathbf{s}(\tau)) \geq 0,\ \ \forall\, \boldsymbol{\sigma} \in K_s^0(s_n(\tau)), \tag{1.35}$$
$$\int_{\Gamma_2} (\tilde{s} - s_n(\tau)) u_n(\mathbf{s}(\tau)) d\Gamma \geq 0,\ \forall \tilde{s} \in -K_n^*, \tag{1.36}$$
$$\mathbf{s}(0) = 0. \tag{1.37}$$

Such a stress formulation has the form of the quasi-variational inequality (1.35) coupled with the variational inequality (1.36), completed by the initial condition. Now the problem arises of how to determine $u_n(\mathbf{s})$. Towards this end the Cesaro method [11] can be used. The classical Cesaro method requires Ω to be simply connected and e_{ij} of class $C^1(\Omega)$. Moreau [10] proposed a generalization to Ω multiply connected and $\mathbf{e} \in \mathcal{D}'(\Omega, E_s^3)$, see also [12]. The latter is the space of distributions. It is worth noting that in the quasi-variational inequality (1.35) the virtual field is the stress field $\boldsymbol{\sigma}$, in contrast with the implicit variational inequality (1.22) where the virtual field is the velocity field.

2. Generalization of the M-CD-M theory of duality

In the static case it is sufficient to consider only one general abstract primal problem P and its dual P^* [16]. The quasi-static case requires investigation of two abstract primal problems here denoted by P and Q, respectively .

2.1. The primal problem P and its dual

The primal problem P involves functionals $\phi(u, v)$, $g(u, v)$, three linear and continuous operators: L, L_1 and L_2 as well as a nonlinear operator N. The space V is a locally convex topological vector space (l.c.t.v.s.), and $(V, V^*, < .,. >)$ is the dual pair. Similarly, $(V_1, V_1^*, < .,. >_1)$ is another dual pair and V_1 is also a l.c.t.v.s. We make the following assumptions:

(H_1) | $w \longrightarrow \phi(v, w)$ is, for each $v \in V$, a convex and lower semicontinous function on V, and $\phi \not\equiv +\infty$.

(H_2) | $v \longrightarrow g(w, v)$ is, for each $w \in V$, a convex function continuous for $v = Lw$.

(H_3a) | N is invertible.

(H_3b) | For each $w \in D(L)$ the mapping $v \rightarrow g(w, v)$ has Gâteaux derivative $D_2 g(w, v)$ with respect to the second variable at $v = Lw$, such that for any $w^* \in V^*$ the set

$$\{w \in D(L) \mid D_2 g(w, Lw) = w^*\},$$

contains at most one element $(D_2 g)^{-1}(w^*)$.

(H_4) | $L : D(L) \rightarrow V \ \ (D(L) \subset V)$, $\ L_1 : V \rightarrow V_1$, $\ L_2 : V^* \rightarrow V_1^*$ are linear and continuous operators. The operator $N : V \rightarrow V^*$, not necessarily linear, is such that $N(0) = 0$.

(H_5) | $\Psi : V_1 \rightarrow R \cup \{+\infty\}$ is a convex, proper lower semicontinuous functional.

Here $D(L)$, $R(L)$ denote domain and range of the operator L. In our case of the Signorini problem with friction we have $L = \frac{d}{d\tau}$ (differentiation in the distributional sense, cf.[9]).

Primal problem P

Find $u(\tau) \in D(L) \cap D(N)$ $(\tau \in [0, \tau_0])$ such that

$$\phi(u(\tau), Lu(\tau)) + g(u(\tau), Lu(\tau)) \leq \phi(u(\tau), v) + g(u(\tau), v), \ \ \forall v \in V, \tag{2.1}$$

$$< L_2 N(u(\tau)), w - L_1 u(\tau) >_1 + \Psi(w) - \Psi(L_1 u(\tau)) \geq 0, \ \ \forall w \in V_1, \tag{2.2}$$

$$u(0) = 0. \tag{2.3}$$

Remark concerning notations. As previously, by $\partial_2 \phi(u(\tau), .)$, etc., we denote the subdifferentiation with respect to the second argument. It means that $\partial_2 \phi(u(\tau), Lu(\tau)) = \partial_2 \phi(u(\tau), v)_{|v = Lu(\tau)}$, and not $L^* \partial_2 \phi(u(\tau), Lu(\tau))$, where L^* is the dual operator of L [17].

Dual problem P^*

Find $u(\tau) \in D(L) \cap D(N)$ and $u^*(\tau) \in V^*$ $(\tau \in [0, \tau_0])$ such that

$$-u^*(\tau) \in \partial_2 g(u(\tau), Lu(\tau)), \tag{2.4}$$

$$\phi^*(u(\tau), u^*(\tau) - < u^*(\tau), Lu(\tau) > \leq$$
$$\leq \phi^*(u(\tau), v^*) - < v^*, Lu(\tau) >, \quad \forall v^* \in V^*, \tag{2.5}$$

$$< -L_1 u(\tau), v_1^* - L_2 u^*(\tau) >_{V_1 \times V_1^*} + \Psi^*(v_1^*) - \Psi^*(L_2 u^*(\tau)) \geq 0, \quad \forall v_1^* \in V_1^*, \tag{2.6}$$

$$u(0) = 0, \quad u^*(0) = 0. \tag{2.7}$$

where $\phi^*(u(\tau), .)$ is the partial Fenchel conjugate of $\phi(u(\tau), .)$, that is

$$\phi^*(u(\tau), v^*) = \sup_{v \in V}\{< v^*, v > -\phi(u(\tau), v)\} \quad v^* \in V^*,$$

and

$$\Psi^*(v_1^*) = \sup_{v_1 \in V_1}\{< v_1^*, v_1 >_1 - \Psi(v_1)\}, \quad v_1^* \in V_1^* .$$

The problems P and P^* are interrelated by

Theorem 2.1. Let the assumptions (H_1), (H_2), (H_4) and (H_5) be satisfied. If $u(\tau) \in V$ $(\tau \in [0, \tau_0])$ is a solution of the primal problem P then there exists $u^*(\tau) \in V^*$ such that $(u(\tau), u^*(\tau))$ is a solution to the dual problem P^*. Conversely, if $(u(\tau), u^*(\tau))$ solves P^* then $u(\tau)$ is a solution of P. Moreover, the following extremality relations hold true

$$\phi(u(\tau), Lu(\tau)) + \phi^*(u(\tau), (u^*(\tau)) =< u^*(\tau), Lu(\tau) >=$$
$$= -[g(u(\tau), L(u(\tau)) + g^*(u(\tau), -u^*(\tau)], \tag{2.8}$$
$$\Psi^*(L_2 u^*(\tau)) + \Psi(L_1 u(\tau)) =< L_2 u^*(\tau), L_1 u(\tau) >_1 . \tag{2.9}$$

Proof.

(i) The variational inequality (2.2) is written in the following equivalent form

$$- L_2 N(u(\tau)) \in \partial\Psi(L_1 u(\tau)) = \partial\Psi(z(\tau))_{z(\tau) = L_1 u(\tau)} \tag{2.10}$$

Hence

$$L_1 u(\tau) \in \partial\Psi^*(-L_2 N(u(\tau)). \tag{2.11}$$

We set $u^*(\tau) = -N(u(\tau))$. Thus (2.11) yields (2.6). Form the assumption (H_4) we infer that $u^*(0) = -N(u(0)) = 0$. The extremality relation (2.9) results readily from (2.10) and Rockafellar's Th.23.5 [13].

(ii) The implicit variational inequality (2.1) is equivalent to the following relations

$$0 \in \partial_2[\phi(u(\tau), Lu(\tau)) + g(u(\tau), Lu(\tau))] = \partial_2(\phi + g)(u(\tau), \; Lu(\tau)) \Longleftrightarrow$$
$$\Longleftrightarrow Lu(\tau) \in \partial_2[\phi(u(\tau), .) + g(u(\tau), .)]^*(0) =$$
$$= \partial_2[\phi^*(u(\tau), .)\nabla g^*(u(\tau), .)](0) =$$
$$= \partial_2\phi^*(u(\tau), (u^*(\tau)) \cap \partial_2 g^*(u(\tau), -u^*(\tau)).$$

where, as we know from the previous step, $u^*(\tau) = -N(u(\tau))$. Here the symbol "∇" denotes the inf-convolution, see [8,13,16]. Now the proof runs similarly as that of Th.2.2 in [16]. To eliminate the primal variable $u(\tau)$ from the dual problem P^* we use either the assumption (H_3a) or (H_3b). Consequently, we shall consider two formulations of the dual problem.

Let us set $N' = -N^{-1}- : v^* \to -N^{-1}(-v^*)$; $D(N') = -R(N)$. If N is linear then $N' = N^{-1}$. Under the assuptions (H_1), (H_2), (H_3a), (H_4) and (H_5) the dual problem P^* is formulated as

Problem P_a^*

$$
\begin{aligned}
&\text{Find } u^*(\tau) \in D(N') \ \ (\tau \in [0, \tau_0]) \ \text{ such that} \\
&-u^*(\tau) \in \partial_2 g[-N'(u^*(\tau)), -LN'(u^*(\tau))], \\
&\phi^*[-N'(u^*(\tau)), u^*(\tau)] - < u^*(\tau), LN'(u^*(\tau)) > \le \\
&\le \phi^*[-N'(u^*(\tau)), v^*] - < v^*, -LN'(u^*(\tau)) >, \ \ \forall v^* \in V^*, \\
&< v_1^* - L_2 u^*(\tau), L_1 N'(u^*(\tau)) >_1 + \Psi^*(v_1^*) - \Psi^*(L_2 u^*(\tau)) \ge 0, \ \ \forall v_1^* \in V_1^*, \\
&u^*(0) = 0.
\end{aligned}
$$

The following theorem interrelates the problems P and P_a^*.

Theorem 2.2. Let the assumptions (H_1), (H_2), (H_3a), (H_4) and (H_5) be satisfied. Then $u(\tau)$ is a solution to the primal problem P if and only if $u^*(\tau) = -N(u(\tau))$ solves the dual problem P_a^*. Moreover, the extremality conditions (2.8) and (2.9) are satisfied.

Remark 2.1. We note that only subdifferentiability of ϕ and g with respect to their second arguments is required by the dual problem P_a^*.

If the condition (H_3b) is satisfied then eliminating the primal variable $u(\tau)$ from the dual problem P, we obtain the dual formulation, now denoted by P_b^*.

Problem P_b^*

$$
\begin{aligned}
&\text{Find } u^*(\tau) \in V^* \ \ (\tau \in [0, \tau_0]) \ \text{ such that} \\
&\phi^*[(D_2 g)^{-1}(-u^*(\tau)), u^*(\tau)] - < u^*(\tau), L[(D_2 g)^{-1}(-u^*(\tau))] > \le \\
&\le \phi^*[(D_2 g)^{-1}(-u^*(\tau)), v^*] - < v^*, L[(D_2 g)^{-1}(-u^*(\tau))] >, \quad \forall v^* \in V^*, && (2.12) \\
&< v_1^* - L_2 u^*(\tau), -L_1 (D_2 g)^{-1}(-u^*(\tau)) >_1 + \\
&+ \Psi^*(v_1^*) - \Psi^*(L u^*(\tau)) \ge 0, \ \ \forall v_1^* \in V_1^*, && (2.13) \\
&u^*(0) = 0. && (2.14)
\end{aligned}
$$

Now we have

Theorem 2.3. Under the conditions (H_1), (H_2), (H_3b), (H_4) and (H_5) an element $u(\tau)$ solves the primal problem P if and only if $u^*(\tau) = -N(u(\tau)) = -D_2 g(u(\tau), Lu(\tau))$ is a solution of the dual problem P_b^*. Moreover, the extremality relations (2.8) and (2.9) hold true.

Let us pass now to the bidual problemP_b^{**}. To obtain its final form we proceed similarly as in [3,16]. Yet, care must be taken because of the operator L.

Problem P_b^{**}

Find $u^*(\tau) \in D(L) \cap D(N)$ $(\tau \in [0, \tau_0])$ such that

$$< D_2 g(u(\tau), Lu(\tau)), v - Lu(\tau) > +\phi(u(\tau), v) - \phi(u(\tau), Lu(\tau)) \geq 0, \quad \forall v \in V, \tag{2.15}$$

$$< L_2 N(u(\tau)), v_1 - L_1 u(\tau) >_1 +\Psi(v_1) - \Psi(L_1 u(\tau)) \geq 0, \quad \forall v_1 \in V_1, \tag{2.16}$$

$$u(0) = 0. \tag{2.17}$$

We have

Theorem 2.4. Let the conditions (H_1), (H_2), $(H_3 b)$, (H_4) and (H_5) be satisfied. A function $u(\tau)$ solves the primal problem P if and only if it is a solution to the bidual problem P_b^{**}.

Remark 2.2. Comparing the problems P and P_b^{**} we infer that they coincide if and only if

$$g(u(\tau), v) =< D_2 g(u(\tau), Lu(\tau)), v - Lu(\tau) > .$$

2.2. Problem Q and its dual

Strees approach leads us to consider an abstract problem Q and its dual Q^*. As previously $(V, V^*, < .,. >)$ is a dual pair, where V is a l.c.t.v.s.

Problem Q

Find $u(\tau) \in V$ $(\tau \in [0, \tau_0])$ such that $\dot{u}(\tau) \in V$ and

$$\phi(u(\tau), u(\tau)) + g(\dot{u}(\tau), u(\tau)) \leq \phi(u(\tau), v) + g(\dot{u}(\tau), v), \quad \forall v \in V, \tag{2.18}$$

$$< L_2 N(u(\tau)), v_1 - L_1 u(\tau) >_1 +\Psi(v_1) - \Psi(L_1 u(\tau)) \geq 0, \quad \forall v_1 \in V_1, \tag{2.19}$$

$$u(0) = 0, \tag{2.20}$$

where $\dot{u}(\tau) = \frac{du}{d\tau}$. The function ϕ satisfies the condition (H_1), while Ψ satisfies the condition (H_5), as previously. Instead of (H_2) we now make the following assumption:

(H_2') | For each $w(\tau) \in V$ the mapping $v(\tau) \longrightarrow g(w(\tau), v(\tau))$ is convex and continuous at $(\dot{v}(\tau), v(\tau))$.

Now we can formulate the dual problem.

Problem Q^*

Find $u(\tau) \in V$ and $u^*(\tau) \in V^*$ $(\tau \in [0, \tau_0])$ such that

$$\dot{u}(\tau) \in V, \quad \dot{u}^*(\tau) \in V^*, \quad \text{and} \quad -\dot{u}^*(\tau) \in \partial_2 g(\dot{u}(\tau), u(\tau)), \tag{2.21}$$

$$\phi^*(u(\tau), \dot{u}^*(\tau)) - < \dot{u}^*(\tau), u(\tau) > \leq \phi^*(u(\tau), v^*) - < v^*, u(\tau) >, \quad \forall v^* \in V^*, \tag{2.22}$$

$$< -L_1 u(\tau), v_1^* - L_2 u^*(\tau) >_{V_1 \times V_1^*} +\Psi^*(v_1^*) - \Psi^*(L_2 u^*(\tau)) \geq 0, \quad \forall v_1^* \in V_1^*, \tag{2.23}$$

$$u(0) = 0, \quad u^*(0) = 0. \tag{2.24}$$

For the sake of simplicity we write $\dot{u}^*$, instead of $\dot{\overline{u^*}}$. In the case considered, subdifferentials $\partial_2 g(\dot{u}(\tau), u(\tau))$, $\partial_2 \phi(u(\tau), u(\tau))$ consist of "velocities". This is a peculiarity of the dual problem. The problems Q and Q^* are interrelated by

Theorem 2.5. Let the conditions (H_1), (H_2'), (H_4) and (H_5) be satisfied. If $u(\tau) \in V$ $(\tau \in [0, \tau_0])$ solves the problem Q then there exists $u^*(\tau) \in V^*$ such that $(u(\tau), u^*(\tau))$ is a solution of the dual problem Q^*. The converse also holds true. Moreover, the following extremality relations are satisfied

$$\begin{aligned}&\phi(u(\tau),u(\tau))+\phi^*(u(\tau),\dot{u}^*(\tau))=<\dot{u}^*(\tau),u(\tau)>=\\&=-[g(\dot{u}(\tau),u(\tau))+g^*(\dot{u}(\tau),-\dot{u}^*(\tau))], \qquad (2.25)\\&\Psi(L_1u(\tau))+\Psi^*(L_2u^*(\tau))=<L_2u^*(\tau),L_1u(\tau)>_1 . \qquad (2.26)\end{aligned}$$

Proof. The implicit variational inequality (2.18) implies

$$0\in\partial_2[\phi(u(\tau),.)+g(\dot{u}(\tau),.)](u(\tau)).$$

Hence

$$\begin{aligned}&u(\tau)\in\partial_2[\phi(u(\tau),.)+g(\dot{u}(\tau),.)]^*(0)=\partial_2[\phi^*(u(\tau),.)\nabla g^*(\dot{u}(\tau),.)](0)=\\&=\partial_2\phi^*(u(\tau),\dot{u}^*(\tau))\cap\partial_2g^*(\dot{u}(\tau),-\dot{u}^*(\tau)), \qquad (2.27)\end{aligned}$$

since $0=u_1^*(\tau)+u_2^*(\tau)$ implies $\dot{u}^*(\tau):=\dot{u}_1^*(\tau)=-\dot{u}_2^*(\tau)$. We take $u^*(\tau)=-N(u(\tau))$. From (2.27) we obtain

$$u(\tau)\in\partial_2\phi^*(u(\tau),\dot{u}(\tau))\Longleftrightarrow\dot{u}^*(\tau)\in\partial_2\phi(u(\tau),u(\tau)), \qquad (2.28)$$

and

$$u(\tau)\in\partial_2g^*(\dot{u}(\tau),-\dot{u}^*(\tau))\Longleftrightarrow-\dot{u}^*(\tau)\in\partial_2g(\dot{u}(\tau),u(\tau)). \qquad (2.29)$$

The relations (2.28) and (2.29) readily yield (2.25). Convexity and lower semicontinuity of the function $\phi(v,.)$ result in

$$\phi(u(\tau),v)=\phi^{**}(u(\tau),v)=\sup\{<v^*,v>-\phi^*(u(\tau),v^*)\mid v^*\in V^*\},\quad v\in V.$$

The last relation and Eq.(2.25) give

$$\begin{aligned}&\phi^*(u(\tau),\dot{u}^*(\tau))-<(\dot{u}^*(\tau),u(\tau)>=\\&=-\phi(u(\tau),u(\tau))\leq\phi^*(u(\tau),v^*)-<v^*,u(\tau)>,\forall v^*\in V^*.\end{aligned}$$

In the last relation we recognize the inequality (2.22).

To eliminate the primal variable from the dual problem Q^* we make the following assumption:

(H_3c) | For each $v^*(\tau)\in V^*$ such that $\dot{\bar{v}}^*(\tau)\in V^*$ and $v^*(0)=0$ the equation $\dot{\bar{v}}^*(\tau)=G_0(\dot{v}(\tau)):=D_2g(\dot{v}(\tau)),v(\tau)$, is uniquely solvable, i.e.: $\dot{v}(\tau)=G_0^{-1}(\dot{\bar{v}}^*(\tau))$. Moreover, we assume that

$$v(\tau)=\int_0^\tau\dot{v}(\xi)d\xi=G_0^{-1}\left(\int_0^\tau\dot{\bar{v}}^*(\xi)d\xi\right)=G_0^{-1}(v^*(\xi)).$$

Taking account of (H_3c) we formulate
Problem Q_c^*

$$\begin{aligned}
&\text{Find } u^*(\tau) \in V^* \quad (\tau \in [0, \tau_0]) \text{ such that } \dot{u}^* \in V^* \text{ and} \\
&\phi^*[G_0^{-1}(-u^*(\tau)), \dot{u}^*(\tau)] - < \dot{u}^*(\tau), G_0^{-1}(-u^*(\tau)) > \leq \\
&\leq \phi^*[G_0^{-1}(-u^*(\tau)), v^*] - < v^*, G_0^{-1}(-u^*(\tau)) >, \quad \forall v^* \in V^*, \qquad (2.30) \\
&< -L_1 G_0^{-1}(-u^*(\tau)), v_1^* - L_2 u^*(\tau) >_{V_1 \times V_1^*} + \Psi^*(v_1^*) - \Psi^*(L_2 u^*(\tau)) \geq 0, \\
&\qquad \forall v_1^* \in V_1^*, \qquad (2.31) \\
&u^*(0) = 0. \qquad (2.32)
\end{aligned}$$

We have
Theorem 2.6. Let the conditions (H_1), (H_2'), (H_3c), (H_4) and (H_5) be satisfied. A function $u(\tau) \in V$ is a solution of the primal problem Q if and only if $u^*(\tau) \in V^*$ solves the dual problem Q_c^*. Moreover, the extremality relations (2.25) and (2.26) are satisfied.

3. Duality for quasi-static Signorini problem with friction

In the present section we derive the dual problem P_d^*, by applying the results of the subsection 2.1. It will also be shown that $P_d^{**} = P_d$, $Q_s^* = P_d$.

3.1. Dual problem P_d^*

We set

$$g(\mathbf{u}, \mathbf{v}) = a(\mathbf{u}, \mathbf{v} - \dot{\mathbf{u}}) - \mathcal{L}(\mathbf{v} - \dot{\mathbf{u}}) - \int_{\Gamma_2} s_n(\mathbf{u})(v_n - \dot{u}_n) d\Gamma, \qquad (3.1)$$

$$\phi(\mathbf{u}, \mathbf{v}) = J(\mathbf{u}, \mathbf{v}), \qquad (3.2)$$

$$\Psi(w) = I_{K_n}(w). \qquad (3.3)$$

The dual problem P_d^* is formulated as a particular case of the problem P_b^*. The relation (3.2) readily yields

$$\phi^*(\mathbf{u}, \mathbf{v}^*) = J^*(\mathbf{u}, \mathbf{v}^*) = I_{C(s_n(\mathbf{u}))}(\mathbf{v}_t^*), \qquad (3.4)$$

where $C(s_n)$ is the (global) set of frictionally asdmissible tangential stresses on Γ_2 for a prescribed normal stress field s_n. For instance, if $s_n \in L^2(\Gamma_2)$ then we have $C(s_n)(x) = K(s_n(x))$, a.e. $x \in \Gamma_2$. Let us find the functional $g^*(\mathbf{u}, \mathbf{v}^*)$. We have

$$\begin{aligned}
g^*(\mathbf{u}, \mathbf{v}^*) &= \sup\{< \mathbf{v}^*, \mathbf{v} > - g(\mathbf{u}, \mathbf{v}) \mid \mathbf{v} \in V\} = \\
&= a(\mathbf{u}, \dot{\mathbf{u}}) - \mathcal{L}(\dot{\mathbf{u}}) - < s_n(\mathbf{u}), \dot{u}_n >_{|\Gamma_2} + \\
&+ \sup_{\mathbf{v} \in V}\{< \mathbf{v}^*, \mathbf{v} > - < B\mathbf{u}, \mathbf{v} > + \mathcal{L}(\mathbf{v}) + < s_n(\mathbf{u}), v_n >_{|\Gamma_2}\}, \qquad (3.5)
\end{aligned}$$

where

$$< B\mathbf{u}, \mathbf{v} >= \int_\Omega a_{ijkl} e_{ij}(\mathbf{u}) e_{kl}(\mathbf{v}) dx = a(\mathbf{u}, \mathbf{v}). \tag{3.6}$$

The linear operator B is invertible, see [16]. We set $G = B^{-1}$. Calculating the supremum on the r.h.s. of (3.5) we finally obtain

$$g^*(\mathbf{u}, \mathbf{v}^*) = a(\mathbf{u}, \dot{\mathbf{u}}) - \mathcal{L}(\dot{\mathbf{u}}) - < s_n(\mathbf{u}), \dot{u}_n >_{|\Gamma_2} + \begin{cases} 0, & \text{if } v^* = B\mathbf{u} - \tilde{\boldsymbol{\mathcal{L}}}, \\ +\infty, & \text{otherwise}, \end{cases} \tag{3.7}$$

where $\tilde{\boldsymbol{\mathcal{L}}} = (\boldsymbol{\mathcal{L}}, s_n(\mathbf{u})) = (\mathbf{B}, \mathbf{F}, s_n(\mathbf{u}))$. The operator G is nothing else than the (static) Green operator. Thus we can write, cf. [11]

$$u_k(\tau) = [G(\mathbf{B}(\tau), \mathbf{F}(\tau), \boldsymbol{\Sigma}(\tau))]_k = \hat{u}_k(\tau) + \int_{\Gamma_2} \Sigma_i(y, \tau) G_{ik}(x, y) d\Gamma(y), \tag{3.8}$$

where $\hat{\mathbf{u}}(\tau) = G(\mathbf{B}(\tau), \mathbf{F}(\tau))$, $\Sigma_i = s_{ij|\Gamma_2} n_j$, $x \in \bar{\Omega}$. In Eq.(3.8) the linearity of the operator G is used.

The Gâteaux derivative $D_2 g$ is calculated from the corresponding differential

$$< D_2 g(\mathbf{u}, \mathbf{v}), \mathbf{h} >= a(\mathbf{u}, \mathbf{h}) - \mathcal{L}(\mathbf{h}) - < s_n(\mathbf{u}), h_n >_{|\Gamma_2}, \tag{3.9}$$

where $\mathbf{h} \in V$. Hence we get

$$D_2 g(\mathbf{u}(\tau), \dot{\mathbf{u}}(\tau)) = B\mathbf{u}(\tau) - \tilde{\boldsymbol{\mathcal{L}}}(\tau) = \boldsymbol{\Sigma}_t(\tau) = \mathbf{s}_t(\tau)_{|\Gamma_2},$$

or

$$B(\mathbf{u}(\tau)) = \tilde{\boldsymbol{\mathcal{L}}}(\tau) + D_2 g(\mathbf{u}(\tau), \dot{\mathbf{u}}(\tau)).$$

Thus we have

$$\mathbf{u}(\tau) = G[D_2 g(\mathbf{u}(\tau), \dot{\mathbf{u}}(\tau))] + G\tilde{\boldsymbol{\mathcal{L}}}(\tau) = G[\boldsymbol{\Sigma}_t(\tau)] + G\tilde{\boldsymbol{\mathcal{L}}}(\tau), \text{ in } \bar{\Omega}.$$

Particularly, on Γ_2 tangential displacements are given by

$$\mathbf{u}_t(\tau) = (D_2 g)^{-1}(\boldsymbol{\Sigma}_t(\tau)) = \hat{\mathbf{u}}_t(\tau) + [G(\boldsymbol{\Sigma}(\tau))]_t \ . \tag{3.10}$$

The last relation results from

$$\begin{aligned} G[\boldsymbol{\Sigma}_t(\tau)] + G[\tilde{\boldsymbol{\mathcal{L}}}(\tau)] = G\boldsymbol{\mathcal{L}}(\tau) + G[s_n(\tau)_{|\Gamma_2}\mathbf{n} + \boldsymbol{\Sigma}_t(\tau)] = \\ = \hat{\mathbf{u}}(\tau) + G[\boldsymbol{\Sigma}(\tau)]. \end{aligned} \tag{3.11}$$

The functional Ψ^* has the simple form

$$\Psi^*(w^*) = I_{K_n^*}(w^*). \tag{3.12}$$

Taking account of (3.4), (3.7) and (3.12) in the abstract dual problem P_b^*, we finally obtain Problem P_d^*

Find $\mathbf{\Sigma}(\tau) = (s_n(\tau), \mathbf{s}_t(\tau)) \in (-K_n^*) \times C(s_n(\tau))$ such that

$$\int_{\Gamma_2} (\mathbf{r} - \mathbf{s}_t(\tau)) \cdot \frac{d}{d\tau}(\hat{\mathbf{u}}_t(\tau) + [G(\mathbf{\Sigma}(\tau))]_t) d\Gamma \geq 0, \quad \forall \mathbf{r} \in C(s_n(\tau)), \tag{3.13}$$

$$\int_{\Gamma_2} (w^* - s_n(\tau))(\hat{u}_n(\tau) + [G(\mathbf{\Sigma}(\tau))]_n) d\Gamma \geq 0, \quad \forall w^* \in -K_n^*, \tag{3.14}$$

$$\mathbf{\Sigma}(0) = 0. \tag{3.15}$$

We see that the dual problem P_d^* consists of the "parabolic" quasi-variational inequality (3.13) coupled with the variational inequality (3.14). It is worth noting that these inequalities are posed on Γ_2 only. By solving the initial-value problem (3.13)-(3.15) we obtain a distribution of normal $s_n(\tau)$ and tangential $\mathbf{s}_t(\tau)$ contact stresses on the surface Γ_2, with $\tau \in [0, \tau_0]$. The extremality relations (2.8) and (2.9) take now the following form

$$J(\mathbf{u}(\tau), \dot{\mathbf{u}}(\tau)) = \int_{\Gamma_2} (-\mathbf{s}_t(\tau)) \cdot \dot{\mathbf{u}}_t(\tau) d\Gamma,$$

$$\int_{\Gamma_2} s_n(\tau) u_n(\tau) d\Gamma = 0.$$

3.2. Proof that $P_d^{**} = P_d$

The proof is straightforward. Taking account of (3.9) we have

$$< D_2 g(\mathbf{u}(\tau), \dot{\mathbf{u}}(\tau)), \mathbf{v} - \dot{\mathbf{u}}(\tau) >= a(\mathbf{u}(\tau), \mathbf{v} - \dot{\mathbf{u}}(\tau)) - \\ -\mathcal{L}(\mathbf{v} - \dot{\mathbf{u}}(\tau)) - \int_{\Gamma_2} s_n(\mathbf{u}(\tau))(v_n - \dot{u}_n(\tau)) d\Gamma.$$

It means that

$$g(\mathbf{u}(\tau), \mathbf{v}) =< D_2 g(\mathbf{u}(\tau), \dot{\mathbf{u}}(\tau)), \mathbf{v} - \dot{\mathbf{u}}(\tau) >,$$

see Remark 2.2. Hence we infer that $P_d^{**} = P_d$.

3.3. Proof that $Q_s^* = P_d$

The stress problem Q_s has the same form as the primal problem Q. In the present subsection we demonstrate that $Q_s^* = P_d$. We set

$$g_1(\dot{\mathbf{s}}(\tau), \mathbf{t}) = b(\dot{\mathbf{s}}(\tau), \mathbf{t}), \quad \mathbf{t} \in H_s, \tag{3.16}$$

$$\phi_1(s_n(\tau), \mathbf{t}) = I_{K_s^0(s_n(\tau))}(\mathbf{t}), \tag{3.17}$$

$$\Psi_1(s) = I_{K_n^*}(s), \quad s \in V_n^*. \tag{3.18}$$

Simple calculation yields

$$g_1^*(\dot{\mathbf{s}}(\tau),\mathbf{t}^*) = \sup\{\int_\Omega t^*_{ij}(x)t_{ij}(x)dx - b(\dot{\mathbf{s}}(\tau),\mathbf{t}) \mid \mathbf{t}\in H_s\} = \\ = \begin{cases} 0, & \text{if } t^*_{ij} = b_{ijkl}\dot{s}_{kl}(\tau), \\ +\infty, & \text{otherwise.}\end{cases} \tag{3.19}$$

We recall that $\dot{\mathbf{s}}$ is the stress rate. For our considerations we may take $\mathbf{t}^* = \mathbf{e}(\dot{\mathbf{v}}),\ \ \dot{\mathbf{v}}\in V$, where V is given by (1.20). The condition (H_3c) is satisfied if elastic moduli a_{ijkl} do not depend on time. Namely we have

$$\dot{\mathbf{s}}^*(\tau) = G_0(\dot{\mathbf{s}}(\tau)) := D_2 g_1(\dot{\mathbf{s}}(\tau),\mathbf{t}) = (b_{ijkl}\dot{s}_{ij}(\tau)). \tag{3.20}$$

Hence

$$G_0^{-1}(\dot{\mathbf{s}}^*(\tau)) = \dot{\mathbf{s}}(\tau) = a_{ijkl}\dot{s}^*_{kl}(\tau). \tag{3.21}$$

where $\mathbf{a} = \mathbf{b}^{-1},\ \ \dot{\mathbf{s}}^*(\tau) = \mathbf{e}(\dot{\mathbf{u}}(\tau)),\ \ \dot{\mathbf{u}}(\tau)\in V$. Integrating (3.20) with respect to time, we get the equation

$$s_{ij}(\tau) = a_{ijkl}e_{kl}(\mathbf{u}(\tau)),$$

provided that a_{ijkl} is time independent and $\mathbf{u}(\tau)$ is continuous at $\tau = 0,\ \ \mathbf{u}(0) = 0$.
Further we have

$$\begin{aligned}
&\phi_1^*(s_n,\mathbf{t}^* = \mathbf{e}(\dot{\mathbf{v}})) = \sup\{\int_\Omega t_{ij}e_{ij}(\dot{\mathbf{v}})dx - \phi_1(s_n,\mathbf{t}) \mid \mathbf{t}\in H_s\} = \\
&= \sup\{-\int_\Omega t_{ij}e_{ij}(\dot{\mathbf{v}})dx \mid \mathbf{t}\in K_s^o(s_n)\} = \\
&= \sup\{-\int_\Omega t_{ij,j}\dot{\mathbf{v}}_i dx + \int_{\Gamma_1\cup\Gamma_2} t_{ij}n_j\dot{v}_i d\Gamma \mid \mathbf{t}\in K_s^o(s_n)\} = \\
&= \mathcal{L}(\dot{\mathbf{v}}) + \tilde{J}(s_n,\dot{\mathbf{v}}_t) + \int_{\Gamma_2} s_n\dot{v}_n d\Gamma,
\end{aligned} \tag{3.22}$$

where

$$\tilde{J}(s_n,\dot{\mathbf{v}}_t) = \sup\{\int_{\Gamma_2}\mathbf{t}_t\cdot\dot{\mathbf{v}}_t d\Gamma \mid \mathbf{t}_t\in C(s_n)\}. \tag{3.23}$$

If $\mathbf{t}_t\cdot\dot{\mathbf{v}}_t\in L^1(\Gamma_2)$ then we have, see [16]

$$\tilde{J}(s_n,\dot{\mathbf{v}}_t) = \int_{\Gamma_2} d(s_n,\dot{\mathbf{v}}_t)d\Gamma\ . \tag{3.24}$$

The inequality (2.30) entering into the problem Q_c^* takes now the following form

$$\begin{aligned}
&-\mathcal{L}(\dot{\mathbf{u}}(\tau)) + \int_{\Gamma_2} s_n(\mathbf{u}(\tau))(-\dot{u}_n(\tau))d\Gamma + \int_{\Gamma_2} d(s_n(\mathbf{u}(\tau),-\dot{\mathbf{u}}_t)d\Gamma + \\
&+a(\mathbf{u}(\tau),\dot{\mathbf{u}}(\tau)) \le -\mathcal{L}(\mathbf{v}) + \int_{\Gamma_2} d(s_n(\mathbf{u}(\tau)),-\mathbf{v}_t)d\Gamma + \\
&+\int_{\Gamma_2} s_n(\mathbf{u}(\tau))(-v_n)d\Gamma + a(\mathbf{u}(\tau),\mathbf{v}),\quad \mathbf{v}\in V.
\end{aligned} \tag{3.25}$$

Further we have $\Psi_1^*(s^*) = I_{K_n}(s^*)$. Thus the variational inequality (2.31) is the same as the inequality (1.23). It has thus been proved that $Q_s^* = P_d$.

The extremality relations (2.25) and (2.26) take now the following form

$$\mathcal{L}(\mathbf{u}(\tau)) - J(\mathbf{u}(\tau), \dot{\mathbf{u}}(\tau)) + \int_{\Gamma_2} s_n(\mathbf{u}(\tau))\dot{u}_n(\tau) d\Gamma = b(\dot{\mathbf{s}}(\tau), \mathbf{s}(\tau)),$$

$$\int_{\Gamma_2} s_n(\mathbf{u}(\tau)) u_n(\tau) d\Gamma = 0,$$

where $\tau \in [0, \tau_0]$.

Concluding remarks

Variational formulation of the quasi-static Signorini problem with friction is available in the form of two coupled inequalities, completed by an initial condition. Both displacement and stress approaches can be used.

Generalizing properly the M-CD-M theory of duality a novel variational problem for the determination of contact normal and tangential stresses has been derived. This variational problem, denoted by P_d^*, consist of the quasi-variational inequality (3.13) coupled with the variational inequality (3.14), completed by the initial condition (3.15). Both inequalities are posed on the surface Γ_2 of possible contact only. The variational problem P_d^* involves (static) Green operator and thus may be interpreted as an alternative formulation of the stress problem Q_s.

In contrast with the static case [16], the complete study of duality for the quasi-static Signorini problem with friction required, on the abstract level, two primal problems denoted by P and Q, respectively. It was shown that in our case the dual problem of the stress problem coincides with the primal one or $Q_s^* = P_d$. It would be interesting to give a physically reasonable example when $Q^* \neq P$.

The generalized M-CD-M theory of duality developed in the present contribution offers new possibilities of application to many frictional contact problem in the presence of the Signorini conditions. Let us mention but a few: quasi-static and dynamic contact problems for two linear elastic solids, frictional contact problem for viscoelastic and creeping solids in the presence of the Signorini conditions.

The general results presented in the paper are by no means restricted to solids. Various structures, like beams, plates and shells can also be examined, provided that they are subjected to frictional and Signorini-like conditions. Many examples of the latter conditions are provided by the paper [15].

References

1. L.-E.Andersson, A global existence result for a quasistatic contact problem with friction, LiTH-MAT-R-89-00, May 1989, Department of Mathematics, Linköping University, Sweden.

2. W.R.Bielski, J.J.Telega, A contribution to contact problems for a class of solids and structures, Arch.Mech., **37**, 1985, 303-320.

3. I.Capuzzo-Dolcetta, M.Matzeu, Duality for implicit variational problems and numerical applications, Numer.Anal. and Optimiz., **2**, 1980, 231-265.

4. I.Ekeland, R.Temam, Convex Analysis and Variational Problems, Elsevier, North-Holland, Amsterdam 1976.

5. N.Kikuchi, J.T.Oden, Contact Problems in Elasticity: A Study of Variational Inequalities and Finite Element Methods, SIAM, Philadelphia, Pennsylvania 1988.

6. A.Klarbring, A.Mikelić, M.Shillor, On friction problems with normal compliance, Nonlinear Anal., TMA, **13**, 1989, 935-955.

7. A.Klarbring, A.Mikelić, M.Shillor, Duality applied to contact problems with friction, in print.

8. P.J.Laurent, Approximation et Optimisation, Herrmann, Paris 1972.

9. J.L.Lions, Quelques Méthodes de Résolution des Problèmes aux Limites Non Linéaires, Dunod, Paris 1969.

10. J.J.Moreau, Champs et distributions de tenseurs déformation sur un ouvert de connexité quelconque, Séminaire d'Analyse Convexe, Montpellier 1976, Exposé No 6.

11. W.Nowacki, Theory of Elasticity, Państwowe Wydawnictwo Naukowe, Warszawa 1970, in Polish.

12. L.Paris, Etude de la régularité d'un champ de vitesses à partir de son tenseur déformation, Séminaire d'Analyse Convexe, Montpellier 1976, Exposé No 12.

13. R.T.Rockafellar, Convex Analysis, Princeton University Press, Princeton 1970.

14. M.J.Sewell, Maximum and Minimum Principles: A Unified Approach with Applications, Cambridge University Press, Cambridge 1987.

15. J.J.Telega, Variational methods in contact problems of mechanics, Uspekhi Mekhaniki (Adv. in Mechanics), **10**, 1987, 3-95, in Russian.

16. J.J.Telega, Topics on unilateral contact problems of elasticity and inelasticity, in: Nonsmooth Mechanics and Applications, ed. by J.J.Moreau and P.D.Panagiotopoulos, pp.341-462, Springer Verlag, Wien-New York 1988.

17. K.Yosida, Functional Analysis, Springer Verlag, Berlin 1978.

J.J.Telega
Polish Academy of Sciences
Institute of Fundamental Technological Research
Świętokrzyska 21
00-049 Warsaw, Poland.

DYNAMICS OF UNILATERAL SYSTEMS ON A FINITE-DIMENSIONAL RIEMANNIAN MANIFOLD[1]

Guo Zhong-heng Wang Shuxun

Department of Mathematics, Peking University

J.J. Moreau in [1-3] discussed the problem of dynamics of unilateral systems of finite degree of freedom, mainly in the framework of Euclidean space. He introduced the concept of soft contact and arrived at the general measure differential inclusion of a nonsmooth motion.

In the present paper the dynamics of uni- and bilateral systems of finite freedom is investigated. Bilateral constraints are eliminated by introducing generalized coordinates. The configuration space is now a finite-dimensional differentiable manifold Q. The assumption of existence of kinetic energy E of the system implies a Riemannian metric on Q. Thus, the original problem reduces to a problem of dynamics of unilateral systems on a Riemannian manifold.

§ 1. Lagrangean Systems and Riemannian Manifold

Let the n-dimensional differentiable manifold Q be the configuration space of a Lagrangean system.

A Lagrangean system may be defined by prescribing the kinetic energy $E: TQ \to \mathbb{R}$ and the generalized force F (covector field on Q), where TQ is the tangent bundle of Q. Let the motion of the system be imaged by the motion of an abstract particle $\mathcal{P}$ along a time-parametric curve

$$C: I \subset \mathbb{R} \to Q: t \to q(t),$$

the tangent vector field $\dot{q}(t) \in Q_{q(t)}$ along it being the generalized velocity.

Let $\{q^i\}$ and $\{q^{i'}\}$ be two local coordinate systems. The kinetic energy E is supposed a time-independent positive definite quadratic form of the velocity $\dot{q}$:

$$E(q,\dot{q}) = 1/2 g_{ij}(q)\, \dot{q}^i \dot{q}^j = 1/2\, g_{i'j'}(q)\, \dot{q}^{i'} \dot{q}^{j'}, \qquad (1.1)$$

[1] Project supported by Science-Technology Foundation of China State Education Commission.

where $g_{ij} = g_{ji}$ (or $g_{i'j'} = g_{j'i'}$) are given functions of q, satisfying the transformation law

$$g_{i'j'} = (\partial q^i/\partial q^{i'})(\partial q^j/\partial q^{j'})g_{ij}.$$

Obviously, we have

Proposition 1.1. The kinetic energy $E : TQ \to \mathbb{R}$ induces a unique second-order covariant tensor field G on Q, defined by the condition:

$$2E(q,\dot{q}) = G(q)(\dot{q},\dot{q}) \qquad \forall q \in Q. \tag{1.2}$$

The positive definiteness of the kinetic energy makes G a Riemannian metric on Q, called the kinetric metric of the system. □

Thus, a Lagrangean system may be characterized by a Riemannian manifold (Q,G) and a covector field F (generalized force) given on it. We shall denote the Lagrangean system by (Q, G, F).

Being considered as a linear mapping, the metric tensor $G(q) : Q_q \to Q^*_q$ is essentially an isomorphism of the tangent space Q_q with cotangent space Q^*_q. Thus, there exists the inverse $G(q)^{-1} : Q^*_q \to Q_q$, which induces a second-order contravariant tensor field H. In a chart $(U, \{q^i\})$ we have the local representations

$$G(q) = g_{ij}(q)dqi \otimes dqj, \tag{1.3}$$

$$H(q) = g^{ij}(q)(\partial/\partial q^i)\otimes(\partial/\partial q^j), \tag{1.4}$$

where $[g^{ij}]$ is the inverse matrix of $[g_{ij}]$.

According to the fundamental theorem of Riemannian geometry [4], there exists a unique torsionless connection D (satisfying $DG = 0$). D is called the Levi-Civita connection on the Riemannian manifold (Q, G). The covariant derivative of the basic vector $\partial/\partial q^j$ along $\partial/\partial q^i$ may be decomposed as

$$D_{\frac{\partial}{\partial q^i}} \frac{\partial}{\partial q^j} = \Gamma^k_{ij} \frac{\partial}{\partial q^k}. \tag{1.5}$$

The connection coefficients $\Gamma^k{}_{ij}$ may be expressed in terms of components of the metric G:

$$\Gamma^k_{ij} = \frac{1}{2} g^{kl} (g_{jl,i} + q_{li,j} - g_{ij,l}), \tag{1.6}$$

where $(\ )_{,k} = \partial/\partial q^k$, and we have

$$D_{\dot{q}(t)}\dot{q}(t) = (\ddot{q}^k + \Gamma^k_{ij}\dot{q}^i\dot{q}^j)\frac{\partial}{\partial q^k}\,. \tag{1.7}$$

Substituting the expression (1.1) for kinetic energy into the local form of the Lagrangean equation

$$\frac{d}{dt}\left(\frac{\partial E}{\partial \dot{q}^i}\right) - \frac{\partial E}{\partial q^i} = F_i\,,$$

we have

$$g_{ij}\,\ddot{q}^j + 1/2\,(g_{ij,k} + g_{ki,j} - g_{jk,i})\,\dot{q}^j\,\dot{q}^k = F_i,$$

or, taking (1.6) into account,

$$g_{il}\,(\ddot{q}^l + \Gamma^l{}_{jk}\dot{q}^j\dot{q}^k) = F_i, \tag{1.8}$$

$$\ddot{q}^i + \Gamma^i{}_{jk}\dot{q}^j\dot{q}^k = q^{ij}\,F_j. \tag{1.9}$$

Taking the expression (1.3,4) into account, we can write (1.8,9) in the global form

$$G(D_{\dot{q}(t)}\dot{q}(t)) = F(q(t)), \tag{1.10}$$

$$D_{\dot{q}(t)}\dot{q}(t) = H(F(q(t)), \tag{1.11}$$

which are independent of local coordinates. Both equations are the Lagrangean equations on the Riemannian manifold $(Q,\ G)$.

The vector field $X = H(F)$ on Q is an alternative representation of the generalized force. In case $X = 0$, the Lagrangean equation reduces to $D_{\dot{q}(t)}\dot{q}(t) = 0$, and the motion $c\colon I \to Q$ is a geodesic on Q.

§ 2. Submanifolds of a Riemannian Manifold

Let M be an embedded submanifold of manifold Q, $i: M \to Q$ being the inclusion map. The Riemannian metric G on Q induces a Riemannian metric i^*G on M. We call (M, i^*G) a submanifold of the Riemannian manifold (Q, G).

The tangent space M_p of M at point $p \in M$ is a subspace of Q_p. In the inner product space $(Q_p, G(p))$, we have the orthogonal decomposition $Q_p = M_p \oplus M_p{}^{\perp}$, $M_p{}^{\perp}$ being the orthogonal complement of M_p, and two projections:

$T : Q_p \to M_p$ (the tangential projection)

$\perp : Q_p \to M_p{}^{\perp}$ (the normal projection)

with

$$X = T X + \perp X \qquad \forall X \in Q_p. \tag{2.1}$$

As the Levi-Civita connection D is induced by G on Q, the induced metric i^*G induces a unique Levi-Civita connection ∇ on M. The following theorem[4] relating these connections is well-known:

Theorem 2.1. If $X_p \in M$, and Y is a tangent vector field on M, then

$$\nabla_{X_p} Y = T(D_{X_p} Y). \quad \square \tag{2.2}$$

Noticing that

$$\perp (D_{X_p} fY) = \perp(f(p)\, D_{X_p} Y + X_p(f) Y_p) = f(p) \perp(D_{X_p} Y),$$

we conclude that there is a well-defined tensor field s, with $s: M_p \times M_p \to M_p{}^{\perp}$ $\forall\, p \in M$, such that

$$s(X_p, Y_p) = \perp(D_{X_p} Y). \tag{2.3}$$

Theorem 2.2. The tensor s is symmetric.

Proof Let X and Y be any extensions of $X_p, Y_p \in M_p$ to all points of Q, which are tangent to M at all points of M. Then

$$\perp(D_{X_p} Y) - \perp(D_{Y_p} X) = \perp(D_{X_p} Y - D_{Y_p} X) = \perp(D_{X_p} Y(p) - D_{Y_p} X(p))$$
$$= \perp([X, Y](p)) = 0,$$

since $[X, Y]$ is also tangent to M at all points of M. $\square$

s is called the second fundamental form of M in Q.

Thus, from (2.1 - 3) we have the Gauss formula

$$D_{X_p} Y = \nabla_{X_p} Y + s(X_p, Y_p) \tag{2.4}$$

Definition 2.3. Let $(M, i*G)$ be a submanifold of (Q, G). The Langrangean system $(M, i*G, T\ X)$ is called a subsystem of the Lagrangean system (Q, G, X).

§ 3. Limit Sets and Parallelism

Definition 3.1. Let $\{A_n\}_{n=1,\dots}$ be a sequence of sets. The limit supremum and limit infimum of $\{A_n\}$ are defined as

$$\overline{\operatorname{Lim}}_{n\to\infty} A_n := \bigcap_{k=1}^{\infty} \bigcup_{n=k}^{\infty} A_n, \tag{3.1}$$

$$\underline{\operatorname{Lim}}_{n\to\infty} A_n := \bigcup_{k=1}^{\infty} \bigcap_{n=k}^{\infty} A_n, \tag{3.2}$$

If these two limits are equal, then

$$\operatorname{Lim}_{n\to\infty} A_n := \overline{\operatorname{Lim}}_{n\to\infty} A_n = \underline{\operatorname{Lim}}_{n\to\infty} A_n \tag{3.3}$$

is called the limit set of $\{A_n\}$.

Definition 3.2. Let X be a tangent vector field defined at each point of a smooth curve $c: I \to Q$ on Q. We say that X is a parallel vector field, if

$$D_{\dot{c}(t)} X(c(t)) = 0 \qquad \forall\ t \in I \qquad \Box \tag{3.4}$$

Let $(U, \{q^i\})$ be a local chart on a neighborhood of point $c(o)$, $\Gamma^k{}_{ij}$ be the connection coefficients, and

$$X(t) = x^i(t) \frac{\partial}{\partial q^i}\Big|_{c(t)},$$

$$\dot{c}(t) = \dot{q}^i(t) \frac{\partial}{\partial q^i}\Big|_{c(t)},$$

then

$$D_{\dot{c}(t)} X(t) = (\dot{x}^k + \Gamma^k_{ij} \dot{q}^i x^j) \frac{\partial}{\partial q^k} .$$

X is parallel along c, iff locally

$$\dot{x}^k + \Gamma^k_{ij} \dot{q}^i x^j = 0. \tag{3.5}$$

The ODE system (3.5) has a unique solution for any initial condition $x^i(o) = x^i{}_0$. In other words, the Levi-Civita connection D of a Riemannian manifold (Q, G) induces a parallelism - a parallel translation operator along any curve c, $P^{c(t)}_{c(t_0)}: Q_{c(t_0)} \to Q_{c(t)}$, such that $P^{c(t)}_{c(t_0)} X_{c(t_0)}$ is a parallel vector field along c. It is easy to show that for any c and $X = \dot{c}(o)$

$$D_X Y = \lim_{t \to 0} \frac{P^{c(0)}_{c(t)} Y(t) - Y(0)}{t}.$$

Proposition 3.3. $X(c(t))$ smooth, iff

$$P^{c(0)}_{c(t)} X(c(t)) : (-\varepsilon, \varepsilon) \to Q_{c(0)} \text{ smooth.}$$

Definition 3.4.

$$\lim_{t \to t_0^-} X(c(t)) := \lim_{t \to t_0^-} P^{c(t_0)}_{c(t)} X(c(t)).$$

§ 4. Unilateral Systems

Roughly speaking, an original Lagrangean system (Q, G, X) becomes a unilateral system if its motion c(t) is additionally confined to

$$c(t) \in \Omega \cup \partial\Omega \qquad \forall t \in I, \tag{4.1}$$

where $\Omega \subset Q$ is an open set and $\partial\Omega$ is its boundary.

In order to give a more precise formulation, we need the notion of tangent cones.

Notation 4.1

$$V_q^+ \Omega := \left\{ \dot{c}(0) \;\middle|\; \begin{matrix} c{:}I \to Q \text{ is a smooth curve} \\ c(0) = q \text{ and } c(0,\varepsilon) \subset \Omega \text{ for some } \varepsilon>0 \end{matrix} \right\} \tag{4.2}$$

$$V_q^- \Omega := \left\{ \dot{c}(0) \;\middle|\; \begin{matrix} c{:}I \to Q \text{ is a smooth curve} \\ c(0) = q \text{ and } c(-\varepsilon, 0) \subset \Omega \text{ for some } \varepsilon>0 \end{matrix} \right\} \tag{4.3}$$

$V^+_q\Omega$ is called the tangent cone of Ω at q. □

It is easy to show

Proposition 4.2.

$$\lambda V_q^{\pm} \Omega \subset V_q^{\pm}\Omega \qquad \text{for } \lambda \geq 0 \tag{4.4}$$

$$V_q^- \Omega = - V_q^+ \Omega . \qquad \square \tag{4.5}$$

Henceforth, $V^{\pm}_q \Omega$ always denotes the closure of $V^{\pm}_q \Omega$ in Q_q. We observe that

$$V_q^+ \Omega = V_q^- \Omega = Q_q \qquad \text{for} \quad q \in \Omega,$$
$$V_q^+ \Omega = V_q^- \Omega = \{0\} \qquad \text{for} \quad q \notin \Omega \cup \partial\Omega.$$

Concerning the boundary $\partial\Omega$, we introduce the following axioms:

(a1) $\partial\Omega$ is can be divided into a finite number of submanifolds:

$$\partial\Omega = (W_{11} \cup \ldots \cup W_{1i_1}) \cup \ldots \cup (W_{n1} \cup \ldots \cup W_{ni_n}). \tag{4.6}$$

Here W_{ij} is a connected regular submanifold of Q with codim $W_{ij} = i$, W_{ij} is maximal in the sense that no other submanifolds from the division (4.6) can join it to form a greater regular submanifold with codimension i, and there exists at least one submanifold W_{st} of higher dimension (s<i) such that W_{ij} is a part of the boundary of W_{st}. (notation $\Omega = W_{01}$)

(a2) $V^+_q\Omega$ is a nontrivial convex cone for every $q \in \partial\Omega$.

(a3) $V^+_q\Omega$ is smooth on every W_{ij}:

$$\lim_{t\to t_0} P_{c(t)}^{c(t_0)} V^{+}_{c(t)}\Omega = V^{+}_{c(t_0)}\Omega \qquad \forall\, c: I \to W_{ij}\,. \tag{4.7}$$

(a4) There is a mapping field A on $\partial\Omega$, such that

$$A_q : V_q^{-}\Omega \to V_q^{+}\Omega \qquad \forall\, q \in \partial\Omega \tag{4.8}$$

satisfies:

(1) A_q is a continuous mapping.

(2) A_q is positive homogeneous:

$$A_q(\lambda u) = \lambda A_q(u), \quad \forall\, \lambda \geq 0 \,;\, u \in V^{-}{}_q\Omega$$

(3) A_q is not energy-producing:

$$\| A_q(u)\| \leq \|u\| \quad , \forall\, u \in V^{-}{}_q\Omega.$$

(4) $A_q|_{V^{-}{}_q\Omega \,\cap\, V^{+}{}_q\Omega}$ = identity, i.e.

$$A_q(u) = u \qquad \forall\, u \in V^{-}{}_q\Omega \cap V^{+}{}_q\Omega.$$

(5) $A|_{W_{ij}}$ is smooth, i.e. A maps smooth sections of the cone bundle $(W_{ij}, V^{-}{}_q\Omega)$ into smooth sections of the cone bundle $(W_{ij}, V^{+}{}_q\Omega)$.

The first three axioms are of geometric character, while the last one is constitutive axiom. A may be called constitutive functor or percussion operator.

Definition 4.3. A unilateral system on the Riemannian manifold is defined as (Q, G, X, Ω, A), where Ω and A satisfy the above 4 axioms.

Henceforth we shall call such systems simply unilateral systems.

§ 5. Motions of a Unilateral System

Definition 5.1. A motion of a unilateral system (Q, G, X, Ω, A) is a mapping

$$c: [0, t_r) \to Q,$$

satisfying

(1) $c(t) \subset \Omega \cup \partial\Omega \qquad \forall\, t \in [0,\, t_r)$,

(2) $\forall t \in [0,\, t_r)$, there exists $\varepsilon > 0$ such that c is smooth on $[t,\, t+\varepsilon]$.

Proposition 5.2. For a motion of the type described above, if $V^+_{c(t)}\Omega$ does not suddenly decrease along $c(t)$, $t \in [0,\, t_r)$, i.e. $\forall\, \tau \in [0, t_r)$

$$\overline{\operatorname{Lim}}_{t \to \tau - 0} P^{c(\tau)}_{c(t)} V^+_{c(t)} \Omega \subseteq V^+_{c(\tau)} \Omega, \tag{5.1}$$

then c is smooth on $[0,\, tr)$.

Proof. By def. 5.1, there exists $\tau \in (0, t_r)$ such that c is smooth on $[0,\, \tau]$. Denote

$$t_s := \sup\{\tau \in (0,\, t_r) \mid c \text{ is smooth on } [0, \tau]\}. \tag{5.2}$$

Suppose that $t_s < t_r$, then we have

$$\dot{c}^-(t_s) = \operatorname{Lim}_{t \to t_s - 0} \dot{c}^+(t) \in \overline{\operatorname{Lim}}_{t \to t_s - 0} V^+_{c(t)} \subseteq V^+_{c(t_s)} \Omega.$$

It means

$$\dot{c}^-(t_s) \in V^-_{c(t_s)} \Omega \cap V^+_{c(t_s)} \Omega.$$

By virtue of condition (4) of (a4), we have

$$\dot{c}^-(t_s) = \dot{c}^+(t_s),$$

and then c is smooth on a neighborhood of t_s, what contradicts the definition (5.2) of t_s. Therefore, $t_s = t_r$. □

The meaning of this proposition is that discontinuity in velocity occurs only when $V^+_{c(t)}\Omega$ decreases suddenly. As a consequence of axiom (a1), we have

Proposition 5.3. For any $q \in W_{ij}$, there exists an open neighborhood $U(q)$ in Q such that $W_{ij} \cap U(q)$ does not intersect any W_{st} of lower dimension $(s > i)$ and belongs to the common boundaries of several $W_{pr} \cap U(q)$ of higher dimension $(p < i)$.

Corollary 5.4. The motion c of the unilateral system has the following local properties at $c(0) \in W_{ij}$:

(R) Right-side property: There exists $\varepsilon > 0$ such that either

(1) $c(t) \in W_{ij}$ $\quad\forall\ t \in [0, \varepsilon)$
or
(2) $c(t) \in W_{pr}$ $\quad\forall\ t \in (0, \varepsilon),\ p < i.$

(L) Left-side property: There exists $\varepsilon > 0$ such that either
(1) $c(t) \in W_{ij}$ $\quad\forall\ t \in (-\varepsilon, 0]$
or
(2) $c(t) \in W_{pr}$ $\quad\forall\ t \in (-\varepsilon, 0),\ p < i.$

Proposition 5.5. For a motion $c\colon (-\varepsilon, 0] \to Q$ with $c(0) = q_0 \in W_{ij}$, we have

$$\lim_{t \to 0^-} P^{c(0)}_{c(t)}\, V^+_{c(t)}\, \Omega \supseteq V^+_{c(0)}\, \Omega. \tag{5.3}$$

Proof. Let $U(q_0)$ be a neighborhood of q_0 in Q and $\Phi : U(q_0) \to \mathbb{R}^n$, $\Phi(q_0) = 0$, be the local coordinate map. $\omega = \Phi\,(\Omega \cap U\,(q_0))$, as an open set in $\mathbb{R}^n$, has a tangent cone at boundary point 0

$$V_0\omega = \Phi_{*q_0}\, V^+_{q_0}\, \Omega.$$

We can approximate the shape of ω near point $0 \in \mathbb{R}^n$ by the convex cone $V_0\omega$. By the convexity of $V_0\omega$, we have

$$a + V_0\omega \subseteq V_0\omega \qquad \forall\ a \in V_0\omega.$$

Thus $V_a\omega \supseteq V_0\omega$ for $a \in V_0\omega$, and then

$$\lim_{t \to 0^-} V_{u(t)}\, \omega \supseteq V_0\omega.$$

Here $u\colon (-\varepsilon, 0] \to \mathbb{R}^n$ with $u(0) = 0$ is the image of curve $c\colon (-\varepsilon, 0] \to Q$ under the mapping Φ. By virtue of the smoothness of the tangent mapping Φ_* and Φ_*^{-1}, we have (5.3).

Theorem 5.6 The motion $c\colon[0, t_r) \to Q$ of the unilateral system (Q, G, X, Ω, A) has the following properties:
(1) The abstract particle $\mathcal{P}$ moves smoothly within any submanifold W_{ij};
(2) When $\mathcal{P}$ moves from submanifold W_{ij} into a submanifold W_{st} of higher dimension $(i > s)$, no percussion occurs and the motion is smooth;

(3) When $\mathcal{P}$ moves from submanifold W_{st} into a submanifold W_{ij} of lower dimension $(i > s)$, percussion occurs and the velocity undergoes discontinuity.

Proof.

(1) As a consequence of axiom (a3), $V^+_q\,\Omega$ is smooth on W_{ij} and the condition (5.1) is fulfilled. Thus c is smooth.

(2) By virtue of the right-side property (R), the condition (5.1) is fulfilled.

(3) By virtue of the left-side property (L), the condition (5.3) is fulfilled, which is opposite to (5.1).

§ 6. Process Chain Method for Cauchy Problem

Let $N_q\Omega$ be the polar (or dual) convex cone of $V^+_q\,\Omega$ in (Q_q, G_q):

$$N_q\,\Omega := \{u \in Q_q \mid G_q(u, v) \leq 0 \qquad \forall\, v \in V^+_q\,\Omega\}. \tag{6.1}$$

Definition 6.1. The unilateral constraint surface $\partial\Omega$ is called ideal, if the constraint reaction R satisfies

$$R(q) \in -N_q\Omega \qquad \forall\, q \in \partial\Omega. \qquad \square \tag{6.2}$$

We shall consider only ideal constraints.

Proposition 6.2. If $T_{ij} : Q_q \to (W_{ij})_q$ is the tangential projection, then

$$T_{ij}R(q) = 0 \qquad \forall q \in W_{ij}. \tag{6.3}$$

Proof. Noticing that

$$(W_{ij})_q^{\perp} = \{u \in Q_q \mid G_q(u,v) = 0 \quad \forall\, v \in (W_{ij})_q\} \tag{6.4}$$

and $(W_{ij})_q \subset V^+_q\Omega$, we have $N_q\,\Omega \subset (W_{ij})_q^{\perp}$, $-N_q\,\Omega \subset (W_{ij})_q^{\perp}$ and $R(q) \in (W_{ij})_q^{\perp}$. Hence, from $R(q) = T_{ij}\,R(q) + \perp_{ij}R(q)$ we have (6.3), because $\perp_{ij}R(q) = R(q)$.

Cauchy Problem Given (Q, G, X, Ω, A) and initial conditions

$$\left.\begin{array}{l} q(0) = q_0 \in \Omega \cup \partial\Omega \\ \dot{q}^{+}(0) = u_0 \in V_{q_0}{}^{+}\Omega \end{array}\right\} \tag{6.5}$$

Determine the motion $q(t)$ for $t \geq 0$.

Solution

Step 1: Without loss of generality, we assume $q_0 \in \Omega$. Solving the Cauchy problem of (Q, G, X) for (6.5), we get the solution $q_1(t)$. Denote the first instant

$$t_1 = \min \{t \,|\, q_1(t) \in \partial\Omega\},$$

when the abstract particle $\mathcal{P}$ percusses $\partial\Omega$ at $q_1(t_1) \in W_{st}$ with input velocity

$$\dot{q}_1^{-}(t) \in V^{-}_{q_1(t_1)}\Omega \quad ,$$

and determine the output velocity

$$\dot{q}_2^{+}(t_1) = A(q_1(t_1))\,(\dot{q}_1^{-}(t_1)) \in V^{+}_{q_1(t_1)}\Omega.$$

Step 2: $\dot{q}_2{}^{+}(t_1)$ is tangent to one or several submanifolds W_{ij} of higher dimension $(0 \leq i \leq s)$. We choose the one with the lowest dimension:

$$\dot{q}_2{}^{+}(t_1) \in (W_{ij})_{q_1(t_1)}.$$

Solving the Cauchy problem of the Langragean subsystem $(W_{ij}, i^{*}G, T_{ij}X)$ for initial conditions

$$\left.\begin{array}{l} q_1(t_1) \in W_{ij} \\ \dot{q}^{+}{}_2(t_1) \end{array}\right\} \tag{6.6}$$

with the Lagrangean equation

$$\nabla^{ij}_{\dot{q}_2(t)}\, \dot{q}_2(t) = T_{ij}\, X, \tag{6.7}$$

we get $q_2(t)$ $\forall\ t \geq t_1$. $q_2(t)$, $t_1 \leq t < t_2$, can be part of the solution of the original Cauchy problem, if it is also a solution of the system $(Q, G, X + R)$ with the equation

$$D_{\dot{q}_2(t)}\ \dot{q}_2(t) = X + R \tag{6.8}$$

for the same initial condition (6.6), provided the constraint force R fulfills condition (6.2).
Substituting (6. 7,8) into Gauss formula, we get the expression for R:

$$R = s_{ij}\ (\dot{q}_2, \dot{q}_2) - \perp_{ij} X. \tag{6.9}$$

In the motion of $\mathcal{P}$ along W_{ij}, there are two possibilities.

Case 1 The particle $\mathcal{P}$ breaks off from W_{ij} and enters into some submanifold W_{pr} of higher dimension $(p < i)$ at the instant

$$t_2 = \sup\ \{t \mid q_2(t) \in W_{ij},\ R\ (q_2(\tau)) \in -N_{q_2(\tau)}\Omega \quad \forall\ t_1 \leq \tau \leq t\}. \tag{6.10}$$

It needs to repeat step 2 for initial condition

$$\left.\begin{aligned} &q_3(t_2) = q_2(t_2) \in W_{pr} \\ &\dot{q}_3^{+}(t_2) = \dot{q}_2^{-}(t_2) \end{aligned}\right\}. \tag{6.11}$$

Case 2 Without detachment $\mathcal{P}$ reaches ∂W_{ij} and enters into some submanifold W_{pr} of lower dimension $(p>i)$ at the instant t_2. Percussion occurs. Instead of $(6.11)_2$ we should use

$$\dot{q}_3^{+}(t_2) = A\ (q_2(t_2))\ (\dot{q}_2^{-}\ (t_2)) \tag{6.12}$$

to repeat step 2. By this way we get a sequence of instants t_3, t_4, ... and the corresponding parts of the solution $q_3(t)$, $q_4(t)$, ...

§7 Examples of operator A

Since (Q_q, G_q) is an inner product space and $V^{*}_{q}\ \Omega \subset Q_q$ is a nonempty closed convex subset, for any $v \in Q_q$, there exists a unique proximal point prox $(v, V^{+}_{q}\,\Omega)$ in $V^{+}_{q}\,\Omega$. Here we give three examples for A which coincide with Moreau's cases:

(1) soft contact $(\delta = 1)$

$$A_1(q)(u^-) = \text{prox}\,(u^-, V^{+}_{q}\,\Omega);$$

(2) elastic percussion $(\delta=0)$

$$A_2\,(q)(u^-) = 2\,\text{prox}\,(u^-, V^{+}_{q}\,\Omega) - u^-;$$

(3) frictionless percussion of class δ

$$A_3(q)(u^-) = \frac{2}{1+\delta}\,\text{prox}\,(u^-, V^{+}_{q}\Omega) - \frac{1-\delta}{1+\delta}\,u^-.$$

References

[1] J.J. Moreau, Standard inelastic shocks and the dynamics of unilateral constraints, in: Unilateral Problems in Structural Analysis, CISM Courses and Lectures No. 288 (ed. G. Del Piero and F. Maceri), Springer, Wien, 1985.

[2] M. Jean, J.J. Moreau, Dynamics in the presence of unilateral contacts and dry friction: A numerical approach, in: Unilateral Problems in Structural Analysis, CISM Courses and Lectures No. 304 (ed. G. Del Piero and F. Maceri), Springer, Wien, 1987.

[3] J.J. Moreau, Unilateral contact and dry friction in finite freedom dynamics, in: Nonsmooth Mechanics and Applications, CISM Courses and Lectures No 302 (ed. J.J. Moreau and P.D. Panagiotopoulos), Springer, Wien, 1988.

[4] M. Spivak, A Comprehensive Introduction to Differential Geometry Vol. III, Publish or Perish, Berkeley, 1979.

Guo Zhong-heng & Wang Shuxun
Department of Mathematics
Peking University
Beijing 100871, China